# BODIES

Geography has recently seen something of a 'body craze'. The politics that surround bodies and spaces are increasingly being held up to scrutiny. Despite this, the 'leaky', 'messy' zones between the inside and outside of bodies and their resulting spatial relationships, remain largely unexamined in the discipline.

This book revolves around three case studies – pregnant bodies in public places, men's bodies in domestic toilets and bathrooms, managers' bodies in Central Business Districts. The pregnant body threatens to expel matter from inside. It is often described as 'ugly' or as 'matter out of place'. Geographers have ignored men's bodies in domestic toilets and bathrooms because these places are abject sights/sites where bodily boundaries are broken and then made solid again. Female and male managers in Central Business Districts wear tailored, dark coloured business suits, that give the appearance of a body which is impervious to leakage or penetration.

The case studies illustrate that bodies and spaces are socially constructed and yet have an undeniable materiality and fluidity. Ignoring the everyday materiality of bodies that 'leak' and 'seep' is not a harmless omission, rather it contains a political imperative that helps keep masculinism intact.

**Robyn Longhurst** is Senior Lecturer of Geography at the University of Waikato, New Zealand.

## CRITICAL GEOGRAPHIES

Edited by

Tracey Skelton, *Lecturer in International Studies, Nottingham Trent University*, and Gill Valentine, *Professor of Geography, The University of Sheffield*

This series offers cutting-edge research organised into four themes: concepts, scale, transformations and work. It is aimed at upper-level undergraduates, research students and academics and will facilitate inter-disciplinary engagement between geography and other social sciences. It provides a forum for the innovative and vibrant debates which span the broad spectrum of this discipline.

1. MIND AND BODY SPACES
Geographies of illness, impairment and disability
*Edited by Ruth Butler and Hester Parr*

2. EMBODIED GEOGRAPHIES
Spaces, bodies and rites of passage
*Edited by Elizabeth Kenworthy Teather*

3. LEISURE/TOURISM GEOGRAPHIES
Practices and geographical knowledge
*Edited by David Crouch*

4. CLUBBING
Dancing, ecstasy, vitality
*Ben Malbon*

5. ENTANGLEMENTS OF POWER
Geographies of domination/resistance
*Edited by Joanne Sharp, Paul Routledge, Chris Philo and Ronan Paddison*

6. DE-CENTRING SEXUALITIES
Politics and representations beyond the metropolis
*Edited by Richard Phillips, Diane Watt and David Shuttleton*

7. GEOPOLITICAL TRADITIONS
A century of political thought
*Edited by Klaus Dodds and David Atkinson*

8. CHILDREN'S GEOGRAPHIES
Playing, living, learning
*Edited by Sarah L. Holloway and Gill Valentine*

9. THINKING SPACE
*Edited by Mike Crang and Nigel Thrift*

10. ANIMAL SPACES, BEASTLY PLACES
New geographies of human–animal relations
*Edited by Chris Philo and Chris Wilbert*

11. BODIES
Exploring fluid boundaries
*Robyn Longhurst*

# BODIES

## Exploring fluid boundaries

*Robyn Longhurst*

London and New York

First published 2001
by Routledge
11 New Fetter Lane, London EC4P 4EE

Simultaneously published in the USA and Canada
by Routledge
29 West 35th Street, New York, NY 10001

*Routledge is an imprint of the Taylor & Francis Group*



Typeset in Perpetua by M Rules
Printed and bound in Great Britain by
TJ International Ltd, Padstow, Cornwall



*British Library Cataloguing in Publication Data*
A catalogue record for this book is available from the British Library

*Library of Congress Cataloging in Publication Data*
Longhurst, Robyn, 1962–
Bodies: exploring fluid boundaries/Robyn Longhurst.
p. cm.– (Critical geographies)
Includes bibliographical references and index
1. Body, Human – Social aspects. 2. Body, Human – Symbolic aspects.
3. Human geography. 4. Feminist theory. I. Title. II. Series.
HM636.L65 2000
306.4–dc21 00-028077

ISBN 0–415–18966–7 (hbk)
ISBN 0–415–18967–5 (pbk)

# CONTENTS

# TABLES

# PLATES

# ACKNOWLEDGEMENTS

Writing this book has in many senses been a collective enterprise. Loved ones, friends, colleagues, peers, students, research participants and acquaintances have all played a role in its production. Consequently, my debts to individuals are too numerous to list in full, but there are many that in memory stand out.

First and foremost I also owe special thanks to Robin Peace. Robin's contributions over the years, as well as to this book, both in the early stages and at the end of the process, have been greatly appreciated. Her personal, collegial and intellectual support has been immeasurable. I am also indebted to Lynda Johnston for her ongoing friendship and astute commentary on a first draft of this book. Throughout the years John Campbell has responded to my numerous questions about 'men's bodies' with a great deal of good humour and intellectual insight. For this I am grateful.

There are many other friends and colleagues who also provided invaluable assistance, albeit at different times, and in different ways. In particular, I would like to thank Richard Bedford, Elaine Bliss, Lex Chalmers, Darryl Gillgren, Lyndell Johns, Catherine Kingfisher, Karen Morin, Karen Nairn, Diana Porteous, Evelyn Stokes, Yvonne Underhill-Sem and Peter Urich. Max Oulton deserves special thanks for his (carto)graphic work. Students in the 1999 graduate geography course 'Crossing boundaries' at the University of Waikato also deserve thanks for engaging in much thought-provoking discussion about 'the body and geography'. Colleen Longhurst, provided fantastic support tending to my children. This space enabled me to write the book and to her I am enormously grateful.

I am also indebted to the research participants who willingly shared their time and experiences for this project. Without these rich narratives the book would not have been possible. Research assistants David Vincent and Marc Elliot played a vital role in collecting interview data and deserve my special appreciation.

I would also like to acknowledge some people whom I have never met but whose ideas have profoundly influenced me, although I hasten to exempt them from any responsibility for how I have used their ideas. In particular I owe a debt to Elizabeth Grosz, Julia Kristeva and Iris Marion Young. I began to read the work of these authors in 1992 when I took part in a course organised by Anna Yeatman and taught by Vicki Kirby in the (what was then entitled) Department of Women's Studies at the University of Waikato. Anna and Vicki fuelled my passion for learning and politics, and the powerful memories of these classes have stayed with me.

I must also thank the editors of the Critical Geography Series, Tracey Skelton and Gill Valentine. The staff at Routledge, especially Ann Michael, ensured that the production process operated smoothly. Thank you.

Thanks also to Arnold for permission to use copyright material for parts of Chapter 2 which appeared as an article in *Progress in Human Geography* (1997) 21, 4: 486–501; Routledge for permission to reproduce some sections of Chapter 3 from E. Teather (ed.) (1999) *Embodied Geographies: Spaces, Bodies and Rites of Passage*; and The Warehouse for permission to use their advertisement for 'Diaper babies'.

An earlier version of parts of Chapter 4 was published in the *Proceedings of the Southern Regional Conference of the International Geographical Union Commission on Gender, Gendered Sites, Human Rights, Gendered Sights, Human Rites*, 8–11 February 1999, University of Otago, Dunedin, New Zealand. Parts of Chapter 5 were published, albeit in a different form, in the *Proceedings of the International Conference on Women in the Asia-Pacific Region: Persons, Powers and Politics*, 11–13 August 1997, National University of Singapore.

Every effort has been made to trace copyright holders, but if any have inadvertently been overlooked they should contact the publishers or me so that the necessary arrangements can be made.

Finally, I dedicate this book to my father, Walter Douglas Longhurst 1915–1988, who always encouraged me to read, think, question and write.

Robyn Longhurst

# 1

# BODILY OPENINGS

This book revolves around three motifs. The first is human geography's[1] disciplinary boundaries. The second is people's bodily[2] boundaries. The third is fluidity – the leaks, flows and filtrations that occur across both disciplinary boundaries and people's bodily boundaries.

In relation to the first motif, the boundaries that organise geographical knowledge into definable categories are discursively[3] unstable. It is not so much that resistance or subversion is needed to overcome these boundaries but rather that the boundaries are subject to constant securing. Despite discursive reiterations disciplinary boundaries are fragile, insecure and increasingly difficult to keep intact now that postmodernist theorists have undermined the self-certainty of the Cartesian subject and questioned claims to unmediated knowledge and truth. Things that are 'loose' are often thought to need securing. Things that are 'secure' always threaten to come loose. This instability of knowledges is, in part, a result of the permeability of discourse itself. Words and meanings are insecure. They can never be pinned firmly on the page.

This book is an attempt to mess up geography's disciplinary boundaries, to scratch the surface of its 'hard' crust, creating interstices through which to slip difference and Otherness[4] into the discipline. It is also about geography's seepage into other disciplines. Focusing on geography's volatile boundaries is a way of questioning some of the discipline's meta-theoretical conceptions. Examining processes of production by identifying some of the ways in which 'we' both disavow and invoke particular kinds of bodies in geographical knowledge might be a useful feminist practice. Knowledge and knowledge production are embodied; they are also sexualised/gendered (Grosz 1993). This leads to the second motif around which this book revolves, people's bodily boundaries.

Over the last few years geographers have begun to pay more attention to bodies – their own bodies (by way of attempts to 'position' the researcher) and the bodies of their research subjects. However, as I argue in this book, the bodies articulated in geographers' texts have tended to be theoretical, discursive, fleshless bodies. A distinction has been drawn between discursive bodies and material bodies. A distinction has also been drawn between political practice and political theory. Discourse and theory seem to offer a purity that materiality and practice threatens to taint and soil. I aim to soil the supposed purity of theoretical possibility by invoking material bodies (this is not to imply that material bodies can simply be plucked from discourse). I want to talk about the shape, depth, biology, insides, outsides and boundaries of bodies placed in particular temporal and spatial contexts. The leaky, messy, awkward zones of the inside/outside of bodies and their resulting spatial relationships remain largely unexamined in geography. This is no accident but rather is linked to a particular politics of masculinist[5] knowledge production.

The third motif around which this book revolves is fluidity – the flows and filtrations that occur across bodies of knowledge and people's bodily boundaries. Luce Irigaray (1985: 113) argues: 'Solid mechanics and rationality have maintained a relationship of very long standing, one against which fluids have never stopped arguing'. Klaus Theweleit (1987a) discusses the desire of the Freikorps World War I German soldiers to stem the tides, floods, streams, waves of the masses and their dread of being engulfed, sucked in, dissolved by the Other. Theweleit does not draw a line between the Freikorpsmen and 'normal' men who also have distaste for fluidity, ambiguity and viscosity. I am interested in a politics of fluidity/solidity in relation to both geography's disciplinary boundaries and people's bodily boundaries. I am also interested in that which is indeterminable between fluid and solid – the viscous. In a review of feminist geography McDowell (1993: 306) refers to work by Adrienne Rich (1986) which suggests that being a woman challenges conventional ideas of boundaries, especially the assumed boundary between the body and the object world, between self and other. McDowell (1993: 306) explains that:

> Women's experiences of, for example, menstruation, childbirth and lactation, all represent challenges to bodily boundaries. The feminine construction of self is an existence centred within a complex relational nexus, compared to the masculine construction of self as separate, distinct and unconnected.

This comment about women's and men's bodily boundaries can be linked to a politics of fluidity/solidity and irrationality/rationality. Women are often understood to be in possession of insecure (leaking, seeping) bodily boundaries; in particular they may leak menstrual blood, and milk from their breasts. It is commonly thought that such bodies are not to be trusted in the public spaces[6] of Rational Man. Men, on the other hand, are often understood to have secure (autonomous) bodily boundaries – bodies that are 'in control'. In some ways these conceptual codings of bodies have little to do with the actual flesh and blood of women and men but in constructing relationships to space they are a powerful force. Over the last decade convincing arguments have emerged about the need to examine new ways of developing frameworks and terms for capturing the multiple, diverse and changing ways in which each human embodied subject is formed. Focusing on a politics of fluidity is one way to do this.

## Mapping the beginning

I first began thinking about bodies, boundaries, and leaks a decade ago when I was contemplating doing doctoral research. In August 1991 I met with the chairperson of the Department of Geography at the University of Waikato in Hamilton, Aotearoa/New Zealand[7] to discuss the possibility. It was the first time we had met but I knew of his reputation for supporting feminist and other critical work. Despite this support, however, he was unable to mask his surprise and disquiet at my suggestion of making bodies the focus of a doctoral dissertation in geography. I explained that I thought bodies play a significant role in people's experiences of place. I wanted to add (but did not) that the sex/gender of bodies – whether we menstruate, ejaculate, have breasts, testicles and so on – affects greatly our relationship with place, and that this is worthy of geographers' attention.

At the time of talking with the chairperson I was a 'lactating mother'. My rapidly changing embodiment over the previous year had often resulted in my exclusion from places where I had previously been welcomed. It had also resulted in my initiation into '(m)other' places of which I knew almost nothing until I became pregnant. I sensed, however, that to proceed with such explanations about the topic of 'the body and geography' meant breaking social (and academic) protocols. I did not feel as though I could use words such as 'ejaculate' or 'lactate' during my first meeting with the geography chairperson. In 1991 (and perhaps still today) there was limited discursive space to discuss bodies, especially their weighty, sexed/gendered and messy materiality in 'respectable' places such as staff offices at the university or in relation to the discipline of geography.

In 1992 I enrolled in the aforementioned geography department to write a doctoral thesis on pregnant bodies in public places. It was a time when many others were also beginning to examine the complex relationship between 'embodiment and spatiality' (Rose 1995: 546). In an article entitled '(Dis)embodied geographies' I outline some of this work and conclude that it is to be welcomed as having 'the potential to prompt new understandings of power, knowledge, and social relationships between people and places' (Longhurst 1997: 496). However, between the time of writing this review and its publication I began to have some niggling doubts about this claim. It was evident that a great deal of work on the body was emerging in geography, but I was no longer sure that it held the emancipatory potential that I had envisaged a few years earlier.

These doubts were fuelled when I took part in two conferences in the United Kingdom. At the 'Cultural Turns/Geographical Turns' conference at Oxford

University (16–18 September 1997) a number of speakers talked about, or at least mentioned, the body. Chris Philo reflected on 'more words' as well as on the need to examine the 'materiality' of things and bodies. John Bale discussed 'Foreign bodies'. Susanne Seymour focused on 'Georgic embodiments of slavery'. Linda McDowell presented an account of 'Embodied workers'. I discussed the need to 'locate' the biology of pregnant bodies. In the plenary session on 'Spaces and subjectivities' all three speakers, Gillian Rose, David Matless and David Sibley referred to the body.

At the beginning of the conference I was excited about this 'development' in geography. I wondered if several decades of feminist scholarship on the body were beginning to pay off. Feminists and other academics interested in critical theory seemed finally to be joining forces to create more 'embodied' and emancipatory geography.

Four months later, 5–8 January 1998, I attended a second conference in the United Kingdom – the Royal Geographical Society-Institute of British Geographers Annual Conference (RGS-IBG) at Kingston. The Social and Cultural Geography Research Group and the Population Geography Research Group had organised an all-day session on 'The body'. Pamela Moss and Isabel Dyck spoke about 'Diagnosing the body: the politics of legitimization'. Tim Cresswell examined 'The mobile body: the construction of the tramp'. Gill Valentine discussed 'Taking up space: the management and discipline of consuming bodies'. And, Rhys Evans spoke about 'Paradise on the margins: (naked) bodies in space at Wreck Beach, British Columbia'. The session was well attended and some excellent papers were presented, but by the end of the day I felt increasingly ambivalent about this turn towards the body. I was uncertain as to whether some of the presentations were necessarily prompting new or 'different' understanding of power.

This may have been a case of resisting what seemed to be suddenly fashionable but more, I think, was at stake. I had a sense that the words 'body' and 'embodiment' were being used at the conferences but often no messy traces of any particular kind of material body were being invoked. Although geographers were using the word 'body' there was often little trace of the specific sexual preference, gender, disability, age, colour, ethnicity or 'race' of bodies. The word 'body' was being deployed, but other words related to the body (for example, muscles, mucus, fart, skin, vulva, sweat, urine, hair, etc.) were rarely mentioned. At both conferences there seemed to be something strangely tidy (and incorporeal) about most of the bodies that were referred to. They appeared to be bodies with secure boundaries, bodies that did not leak or seep, bodies that were clean, mess- and matter-free.

The experience of writing a doctoral dissertation on 'Pregnant bodies in public spaces' and of attending the aforementioned conferences in the United Kingdom prompted me further to examine this thing called 'the body' and some of the implications of geographers 'placing' it more centrally in our work. I am no longer sure that some of the current work on the body (and I include my own here) offers routes to achieving more emancipatory social and political relations. As a result of these misgivings I began to think about bodily boundaries – the liminal places where the exteriority and interiority of bodies merge. I also began to think about disciplinary boundaries and that examining bodily boundaries might be a way of attending to geography's disciplinary boundaries.

There is little doubt that geographers in the 1990s have considered and invoked the body in their/our work, but exactly what kind of body is being invoked and what is at stake in such a move? Does this 'new' turn hold any kind of emancipatory potential for those interested in addressing issues of social justice, in particular, rights for women? Does focusing on the body fundamentally alter in any way the ontological and epistemological foundations of human geography? Exactly what kinds of bodies are being invoked in geographical discourse? These are the questions that prompted me to write this book. They are difficult questions to answer and therefore I have proceeded with caution. In the final instance I am uncertain as to exactly how content I am with my response to these questions. My hope is that what I have written will prompt others to engage with these ideas and subsequently put forward their views and perspectives.

## Charting the journey

Chapter 2 begins with an argument that bodies are 'real' (have a weighty materiality), while at the same time, they are socially constructed (are enmeshed in discourses). Bodies are also always in a state of becoming with places. This point might appear obvious to those who are interested in thinking about space and place, but it is a point that is still frequently overlooked by many social theorists. The question 'what is a body?' can only be answered by 'locating' bodies. I trace the 'absent presence' (Shilling 1993), the Othering, of the body in geography over the last three decades. Underpinning this chapter is the question: 'what are some of the epistemological and ontological implications of focusing on the body for the discipline of geography?' There is now a great deal of research in which the body is being invoked in geography but is this research altering some of the disciplines' meta-foundations? If so, in what ways? Does focusing on the body offer a way of prompting new understandings of power, knowledge and social relationships between people and places? Chapter 2, and the book as a whole, is intended as a starting rather than a finishing point for a discussion on bodies and geographical epistemologies and ontologies. In the chapters that follow I move from the theoretical to the empirical and focus on specific bodies in specific times and places.

Chapter 3 focuses on pregnant women in public places. When pregnant women occupy the public spaces (of Rational Man) they are often represented as 'matter out of place' (Douglas 1966: 35). Women are often thought to threaten and disrupt a social system that requires them to remain largely confined to private space during pregnancy. Pregnant women can be seen to occupy a borderline state that disturbs identity, system and order by not respecting borders, positions and rules. The pregnant body, it is thought, threatens to expel matter from inside – to seep and leak. The pregnant woman may vomit (morning sickness), cry (she is constructed as 'overly' emotional), need to urinate more frequently, produce colostrum which may leak from her breasts, have a 'show' appear, have her 'waters break', and sweat with the effort of carrying the extra weight of her body. Even more than these leakages, she 'threatens' to split her one self into two – another human being is about to cross the boundary of the 'eroticised orifice' – the vagina (Grosz 1990: 88). The pregnant body is neither subject nor object but rather exemplifies the impossible, ambiguous and untenable identity of each. Consequently, the pregnant body is often constructed as abject. It is a body that is considered dangerous and to be feared. It is also considered to be a body that needs to be controlled.

There are many ways in which attempts are made to control pregnant bodies. First, the fetus is often treated as though it were a public concern. Pregnant women's rights to bodily autonomy are considered to be questionable. Second, this leads to pregnant women's stomachs being subject to public gaze and often touch. Their 'bodily space' is frequently invaded. Third, pregnant women tend to be constructed by lovers, husbands, partners, friends, family, strangers, health workers and themselves as being in a 'condition' in which they must take special care in order to protect the well-being of the fetus. Fourth, pregnant women are subject to dietary regimes in an attempt to control what enters their bodies.

In Chapter 4 the focus shifts from women's bodies to men's bodies. It also shifts from bodies that tend to be constructed and understood as 'vulnerable' (pregnant women) to bodies that tend to be constructed and understood as hard and strong (heterosexual, 'white', able-bodied men). The focus also shifts from public space to private space, in particular, domestic toilets/bathrooms. These rooms are often the site where bodily boundaries are broken and then made solid again. The body is (re)made ready for public scrutiny. I examine men's experiences of domestic toilets/bathrooms as a strategic move to displace the often implicit association of body fluids and flows with femininity (especially maternity, pregnancy and menstruation).

Several themes are discussed in this chapter. The first is men's fear of (homosexual) contamination by 'circuits of fluid'. This includes disquiet about seminal fluid being left in toilets/bathrooms and reluctance by men to share body products because they may inadvertently share bodily fluids. The second theme is the construction of domestic bathrooms as feminised spaces. Many men avoid lingering in bathrooms, treating them a functional space rather than a space of a pleasure. The third theme is men's discomfort with reminders of (the residues of) their corporeality. This includes 'dirty' bath water and excrement that will not flush. This chapter concludes with a note about the exclusion of toilets/bathrooms from geographical discourse. Abject sights/sites such as toilets/bathrooms and the bodies that inhabit them threaten to mess up clean 'hard' masculinist geography thus indicating some of the limits about what we can do and say in the discipline.

Ideas about body image and presentation are also pursued in Chapter 5, but in this chapter the focus returns to three aspects of managerial bodies in public space, particularly men and women managers who work in central business districts (CBDs). I argue, first, that highly tailored, dark-coloured (often black, dark grey or navy) business suits function to seal the bodies of (men and women) managers. Firm, straight lines and starched creases give the appearance of a body that is impervious to outside penetration. They also give the appearance of a body that is impervious to the dangers and threats of matter from inside the body making its way to the outside. It is considered inappropriate for matter to make its way from the inside to the outside of bodies (for example, farting, burping, urinating, spitting, dribbling, sneezing, coughing, having a 'runny nose', crying, and sweating) in most inner city workplaces. This suited, professional, respectable body, however, can never be guaranteed. Like all bodies it is continually monitored and disciplined but inevitably proves itself to be insecure.

Second, I discuss the disciplining of employees' bodies. Although the prescriptions for the ideal presentation of employees varies somewhat, all managers play an explicit role in disciplining the bodies of 'their' workers. Managers shape not only their own 'look' but also the 'look' of their employees. Just as important as the clothes, however, is the flesh of the body itself – the comportment, gait, fat, hair, muscles, skin tone and colour and facial features.

This provides the third focal point. Managers are currently being expected to 'live out' discourses of flexibility at the level of the flesh. Their bodies are being disciplined (both by the self and by others) to look 'good' and appear fit and flexible at work. This discourse functions in different ways for men and women. A new kind of masculinity – one that requires men to be trim, taut and terrific – is emerging. This hegemonic construction of masculinity functions to disadvantage many groups including women, the disabled, people who are large, elderly, unfit, fragile, ill and/or 'disfigured' with birth marks, acne or scars.

In the concluding chapter I evaluate what it means to put bodies (bodies that are inseparable from minds, bodies that have specificity and uniqueness in terms of features, colour, genitalia, size and age, and bodies that break their boundaries) centre-stage in geography. I return to discuss pregnant bodies in public spaces, men's insecure bodies in toilets/bathrooms, and managers' 'suited', fit and flexible bodies in CBDs in order to reflect on body–space relations. All three examples illustrate that bodies and spaces are socially constructed and yet have an undeniable materiality. They also illustrate that bodies and spaces are neither clearly separable nor stable. For example, the interuterine space of the pregnant body, the defecating man in the toilet, and the manager whose body inevitably fails to be respectable all the time, illustrates not only a close(t) relationship between bodies and spaces but also a leaking and inseparability of bodies and spaces. The spaces of bodies themselves and the spaces of places do not remain clearly separable but make each other in everyday ways. The book ends with a question about what it means to write from 'down under', the South, the Antipodes, New Zealand (see Peace, Longhurst, Johnston 1997 on 'producing feminist geography "down under"'; see also Berg 1994; Berg and Kearns 1996, 1998; Mohanram 1999). My question is: what part has my 'spatial identity' played in the creation of this text about geographical and bodily boundaries?

Although each chapter in this book revolves around disciplinary and bodily boundaries, abjection and close(t) spaces, there is no clear linear argument that runs throughout. Chapters can be read as individual essays or the book can be read from cover to cover. Whichever is your preference, my hope is that you will be prompted to question further some of the grounds upon which knowledges about bodies and spaces rests.

# 2

# 'CORPOREOGRAPHIES'[1]

An advertising slogan for Magic Diaper Babies claims that the purchaser can: 'Put the baby in cold water to discover whether it is a boy (blue) or a girl (pink)' (see Plate 2.1). When such advertisements circulate through household letter boxes (as they did in New Zealand in 1992) a symbolic economy is invoked that would be understood by most people in the 'western' world. In fact the urine of 'real' male and female infants is the same colour – opaque yellow. Urine is not (usually) blue or pink but a symbolic economy operates which means that most people in the western world understand this advertisement and would not stop, even momentarily, to puzzle over its meaning. The urine is used to 'gender' the doll.

An advertisement for sanitary napkins (pads) also circulating through New Zealand letter boxes showed blue liquid being poured on and seeping into the pad. The aim seemed to be to demonstrate the pad's absorbency. Similar advertisements have been, and continue to be, screened on television in many countries. Although

*Plate 2.1* Diaper babies.
*Source:* The Warehouse advertising flier, June 1992.

the pad is designed to absorb menstrual blood (various shades of red) the advertisers choose to use a blue liquid to illustrate the product's absorbency capacity. Readers of this advertisement are not supposed to understand women to be gendered male (the meaning implied by the blue urine of the baby doll), rather, a different symbolic order is invoked. While the aim of the Magic Diaper Babies is presumably to pique children's curiosity about sexual difference, the aim of the pads is to keep menstrual flow 'in order'.

These advertisements placed side by side illustrate a complex discourse about sexual difference and bodily fluids. If advertisers were to use red fluid to represent blood they might invoke in viewers and potential customers feelings of dirtiness, disgust and even death that would dissuade them from purchasing the product. Clear blue liquid, on the other hand, is often associated with purity and cleanliness. There are potential links with running water, the colossal cleaning capacity of the ocean, blue granules in cleaning products and laundry powers, and blue rinses (both for clothes and for (often wealthy) elderly women's greying hair). It is also inevitable that a connection may be made with the term 'blue blood'. Blue blood refers to

> that which flows in the veins of old and aristocratic families . . . who claimed never to have been contaminated by Moorish, Jewish, or other foreign admixture; the expression probably originated in the blueness of the veins of the people of fair complexion as compared with those of dark skin.
>
> (*Compact Oxford English Dictionary* 1991: 147)

On all accounts blue blood seems to be preferred over and above red blood.

There is a point to examining these seemingly banal advertisements. Vicki Kirby (1997) disturbs dualistic[2] categories such as mind/body, culture/nature and signification/flesh by suggesting that cultural contexts do not simply surround bodies but also come to inhabit them. In this case the advertisements for baby dolls or sanitary napkins do not simply envelop or rewrite the same universal bodies, rather, they reconstitute the matter of the body – the skin, organs and cells – in a myriad of complex ways. Kirby argues that bodies are not just texts written upon by representational fields, they are flesh articulated by language (of which there is no outside). This understanding of corporeality means that advertisements such as the examples cited above are highly significant in considering relationships between place, politics and identity.

I begin this chapter by addressing the question 'what is a body?' arguing that the matter of the body cannot be taken for granted and treated as obvious. The term 'body' cannot easily be contained with a neat dictionary definition or a commonplace understanding of what it means. In order to understand bodies it is necessary to pay attention to discourses and/on/in flesh. Second, I examine some of the ways in which the body has been Othered in particular schools of geographical 'thought' (including positivist, humanist, Marxist and feminist geographies) in the 1960s, 1970s and 1980s. The third point to be argued is that the 1990s has seen an upsurge of interest in the body and the question is raised: 'why now, this turn towards the corporeal?' Fourth, some of the flurry of work that focuses on bodies and spaces is outlined, including conference sessions, research publications and teaching. In the penultimate section I argue that although many geographers are now paying attention to the body it is still not acceptable for the flesh and boundaries of fluid, volatile, messy, leaky bodies to be narrated. Finally, I point to the notion of abjection suggesting that it is potentially useful for geographers interested in considering the corpo(real) and dislodging the masculinism constructed in bodies assumed to be autonomous and rational.

## What is a body?

> What's inside a body? More bodies. Body wholes, body parts, body fluids. We are already inside-out but we don't yet know it. We still can't recognise our bodies in ourselves.
>
> (Brophy 1988: storyboard used in the film *Salt, Saliva, Sperm and Sweat*)

Phillip Brophy, using a 'storyboard' (represented as text on a computer screen) in a film that takes viewers on an extraordinary odyssey into the banalities of everyday life, poses the question 'what's inside a body?' but not 'what *is* a body?' The matter of the body encompasses a 'bewildering variety' of meanings: it is 'equivocal, often ambiguous, sometimes evasive and always contested' by those who attempt to understand its meaning more fully (Pile and Thrift 1995: 6). We all have one, or at least, we all are one. We are all born, we all die. Although these things appear to be universal our embodied experiences are unique (Nast and Pile 1998: 1). Our bodies carry out a vast array of tasks and pleasures. They can be surfaces for artwork, carry out practical tasks, establish and cement identities, make love, violate, excrete and pray (see Harré 1991: 257 cited in Pile and Thrift 1995: 6). Bodies and their socially encoded meanings can be understood only in specific spatial, temporal and cultural contexts.

There has been recent debate on the body,[3] yet, the seemingly simple question 'what is a body?' has not been examined thoroughly. Those theorists who do attempt to address the question often remain puzzled. Bryan Turner (1984: 7 cited in Kirby 1992: 1) admits that at the end of writing his book *The Body and Society*, he was even more confounded by the '"crassly obvious" question "what is the body?"' than when he began. Vicki Kirby (1992: 1) probes this puzzling matter commonly called the body and claims that it is 'a *terra incognita*'. She wonders, how we might *think* this 'corporeal place'. Elizabeth Grosz (1992: 243), who for a number of years has researched embodiment, claims:

> By *body* I understand a concrete, material, animate organization of flesh, organs, nerves, muscles, and skeletal structure which are given a unity, cohesiveness, and organization only through their psychical and social inscription as the surface and raw materials of an integrated and cohesive totality . . . The body becomes a *human* body, a body which coincides with the 'shape' and space of a psyche, a body whose epidermic surface bounds a psychical unity, a body which thereby defines the limits of experience and subjectivity, in psychoanalytic terms through the intervention of the (m)other, and ultimately, the Other or Symbolic order (language and rule-governed social order) [emphasis in original].

Grosz's definition allows us some sense of what bodies might be but the 'matter' 'at hand' remains problematic. Clearly it is impossible, and not necessarily very useful, to attempt to offer any kind of absolute or exact definition of the term. Grosz's definition provides some useful, initial explanation of this thing called a body.

A field of study that recognises bodies as a way of understanding the relationships between people and places is emerging in geography. When I began a doctoral thesis on pregnant bodies in 1992 there was little geographical research on the body. My project was subject to ridicule from colleagues and the popular press. One of my colleagues thought it 'absolutely disgusting' that I should examine the provision of public toilets for pregnant women. The editors of an Auckland-based, glossy, current affairs magazine, *Metro*, ridiculed an advertisement for a presentation to a branch of the New Zealand Geographical Society on 'Pregnant corporeality and a postmodern mall'. The advertisement was duplicated on the last page of *Metro* (1993: 138) under the editorial heading 'Wankers' World'. In the masturbatory geographical imaginary of the editor's mind the connection between geography and corporeality, or perhaps between geography and pregnancy, or between architecture and pregnancy, was supposed to be absurd, ridiculous and humorous. The 'world' of the university, especially faculties of arts and social sciences, is a world of wankers. Anti-intellectual sentiments tend to run high in New Zealand (see Berg 1994; Longhurst 1994a; see also Bell 1995 on censorship practices in geography).

One explanation for derogatory comments about work on the body stems from arguments about the mind and body dualism. Moira Gatens (1988: 61) claims: 'not only have mind and body been conceptualised as distinct in western knowledges but also the divisions have been conceptually and historically sexualised'. Genevieve Lloyd (1993) examines the works of various philosophers (for example, Plato,

Aristotle, Bacon, Philo, Augustine, Aquinas, Descartes, Hume, Rousseau, Kant, Hegel, Satre and de Beauvoir) in order to trace associations between ideals of human reason and ideals of masculinity. One of the main points Lloyd makes is that a form/matter or mind/body distinction operated, although in different ways at different times, in Greek, and subsequent theories, of knowledge.

Margrit Shildrick (1997) discusses conceptualisations of the mind and body through various historical periods making specific reference to medical discourses. For example, Shildrick refers to the 'Renaissance anatomists' (p. 15), the 'Vesalian drawings in the "Epitome" *of De humani corporis fabrica* (1543)' (p. 18) and 'Galen's dissections not of humans but of animals' (p. 20). Grosz (1989: xiv) argues that the mind has traditionally been correlated with positive terms such as 'reason, subject, consciousness, interiority, activity and masculinity'. The body on the other hand, has been implicitly associated with negative terms such as 'passion, object, non-consciousness, exteriority, passivity and femininity' (ibid.). The body has been seen as reason's 'underside', its 'negative, inverted double' (Grosz 1988: 30). Numerous feminist theorists have convincingly argued that such dualisms are gendered/sexed (see Bordo 1986; Jay 1981; Le Doeuff 1987, 1991; Lloyd 1993). Dualistic thinking has also been heavily criticised by postmodern theorists.[4]

Of course, in 'reality', both men and women 'have' (or rather, 'are') bodies. The difference is that men are widely considered to be able to seek and speak universal knowledge, unencumbered by the limitations of a material body placed in a particular material context whereas women are thought to be bound to the desires of their fleshy, 'natural' bodies placed in time and space. In western culture, while white men may have presumed that they could transcend their embodiment (or at least have their bodily needs met by others) by seeing the body as little more than a container for the pure consciousness it held inside, this was not allowed for women, blacks, homosexuals, people with disabilities, the elderly and children. This masculinist separation of minds from bodies, and the privileging of minds over bodies, remains a dominant conception in western culture although it has been challenged.

## Other(ed) bodies

Over the last three decades bodies have been constructed as Other in geography (Longhurst 1997). They have functioned as a lesser category to the mind and that which has long been associated with it – public space, rationality and objectivity. Rose (1993a: 6) claims that reason (and one could add here, the mind) 'is not the whole story of masculinism . . . in order to establish rationality, there must be a contrast with the irrational' (and one could add here, the body). 'Disciplinary knowledge can define itself through its own ability to know only if there are others who are incapable of knowing. For a masculinity defined in part through its rationality, its Other must be deemed irrational' (Rose 1993a: 9). Geography has been premised on a Cartesian dualism between mind and body (Longhurst 1997).

The exact ways in which bodies have functioned as Other in geographical discourse depends on the particular school of thought under consideration. In the 1960s and early 1970s when geography was arguably at a peak of embracing positivist perspectives ('factual' statements or hypotheses verified through empirical testing) the body functioned as Other to the mind. An assumption, which underlies positivism, is that the geographer is an objective and disembodied observer, detached from the subject being studied. In the 1960s and early 1970s geography was understood by many to be a 'spatial science'[5] but even during this period the body was not entirely absent.

In the most seemingly disembodied of geographical texts, the body leaves its trace. For example, in B. H. Hodder and Roger Lee's (1974) *Economic Geography*, the body appears to give way almost totally to discourses of 'decision and control', 'logical discussion', 'efficiency', 'models' and 'capital'. Yet on the opening page Hodder writes 'For my Parents'; Lee writes 'For Lesley; Thomas, who embodies all things good in my life; and for my Mother'. The language of rationality and disembodiment cannot altogether suppress that which is irrational, passionate and embodied. Such traces of the irrational, the emotional, the domestic and the body are often evidenced on the acknowledgement page of otherwise rationalistic texts.

In a sense it is easy to track the body through the acknowledgement page but somewhat different traces of the body can also be found in the texts themselves. In discussing differences in consumer behaviour Hodder and Lee (1974: 53) claim that these differences cannot be ascribed simply to differences in household size or income: 'non-monetary variables associated with household composition, for example age and sex structure and stage in the life-cycle' must also be considered. Objective economic analysis falters and gives way, at least for one page, to social and psychological studies of consumer behaviour. Terms such as economics, profits, markets and incomes are supplemented, at least briefly, with terms such as stage in the life-cycle, personal filters, cultural status and attitude formation. The authors acknowledge that differences in consumer behaviour between households cannot be explained by totally rationalist discourse. In other words, traces of the body, the emotional and irrational, do exist in the text, although they remain little more than traces. The body in geographical literatures such as Hodder and Lee's (1974) text is not made explicit – it is present but functions as a backdrop to geographers' analyses rather than as the subject in its own right.

In the late 1970s the body began to function in different ways in geographical research and texts. The late 1970s saw the rise of humanistic perspectives in geography. These perspectives, in contrast to positivist perspectives, emphasise the subjectivity of both the geographer and the observed phenomena. Humanistic geographers (such as Buttimer 1976, 1979; Ley and Samuels 1978; Relph 1976, 1981;

Tuan 1976) claim that in order to understand and interpret social action it is necessary to do more than just analyse data. Prolonged contact is needed with subjects in order to appreciate how they learn, act and react, how they obtain, process and transmit information, and how they assign meaning to people and places.[6]

Humanistic geographers attempted to understand the ways in which places were perceived by people, arguing that it was impossible to make sense of the social world without paying attention to the interpretations of those who lived in it (see Buttimer 1979; Entrikin 1976; Ley and Samuels 1978; Relph 1976; Tuan 1976, 1979). In this way, people were placed centre-stage. Drawing on phenomenology, they focused not on reason but on human creativity, not on the universal but on the specifics of people's lives. Rose (1993a: 48) argues that by drawing from phenomenology humanistic geographers retrieved the Other – in this case, the body – into their studies. For example: '"Body implicates space; space co-exists with the sentient body", wrote Tuan [1974, 218]. Topophilia is felt in part corporeally' (Rose 1993a: 48).[7]

Yi Fu Tuan was not the only humanistic geographer to bring the body into geography. In the late 1970s David Seamon used the pre-discursive, phenomenological lived body, theorised in the work of Merleau-Ponty, to explore some of the ways in which people move through and occupy space. In particular he observed and carried out interviews with groups of students in order to elicit dimensions of their life-worlds (see Johnson 1989).

Grahame Rowles (1978a, 1978b) also examined the lived geographies and life-worlds of his subjects and demonstrated that the body, both real and imagined, actively constructs space. 'Rowles's engagement with elderly people led him to recognise the significance of immobility and fantasy displacements to other places' (Johnson 1990: 18). In this way, humanistic geographers made bodies far more visible in geography, highlighting the need to consider embodied subjectivities for understanding place. Rose (1993a: 48) explains that: 'Analysis and empathy, insider and outsider, thought and pleasure, body and mind, individual and context; these are some of the dualisms that humanistic geography explicitly attempted to overcome in its efforts to interpret the world'.

Nevertheless, while corporeality played an important role in humanistic geography in the 1970s, the body still tended to function as an Other in a number of ways. First, humanistic geographers failed to acknowledge the *specificity* of bodies, for example, peoples' various skin colours, shapes, abilities, features and sex/gender. In relation to sex/gender Seamon failed to recognise that observable differences exist in the rather ordinary ways in which women and men typically access, move in, use, perceive and experience environments. While he may have succeeded in embodying his subjects, he did not recognise the significance of these bodies being *sexed*. This is problematic since, as Gatens (1991a: 82) explains, references to and representations of the *human* body are most often of the *male* body. In the absence of any particular body being specified, a white, masculine, self-contained body is presumed (see Rose 1993a).

Humanist geographers David Ley and Marwyn Samuels (1978: 2–3 cited in Rose 1993a: 43) wanted 'man put back together again with all the pieces in place, including a heart and even a soul, with feelings as well as thoughts, with some semblance of secular and perhaps transcendental meaning'. The body they refer to here is, I suspect, indeed a *man's* body. Ley and Samuels desire a body that is whole, complete, autonomous, transcendent and almost incorporeal – they want a body that is 'put back together with all the pieces in place'. They do not want a body that is messy, incomplete, out of place and not possessing clear boundaries. They do not argue for the menstruating, birthing or lactating body – that which is associated with the feminine. Thus, although humanistic geography at this time included the body in its discourse its (able-bodied, white) masculinism remained intact. The use of Man in humanistic geography helped to make men the base-line against which Woman's embodied difference was spoken. 'The authority of humanistic geography is masculinist because it falsely assumes that the experiences [including the bodily experiences] of men can represent all experiences' (Rose 1993a: 53).

Following the publication of David Harvey's *Social Justice and the City* in 1973, Marxist approaches to geography also began to emerge in the early 1970s. The potential of such approaches, however, was not fully realised until the 1980s when a number of geographers (such as Dear and Scott 1981; Johnston 1984; Short 1985; and Smith 1984) began to engage with Marxist political economic theory in relation to spatial configurations. Marxist geographers were 'concerned with the ways in which the production of space, place and landscape is implicated in the reproduction of specific social formations' (Johnston, Gregory and Smith 1986: 287). A new discourse emerged in geography – a discourse that included terms such as modes of production, class struggle (the bourgeoisie and the proletariat), labour power, base and superstructure, (re)production, and exchange/surplus/use value. It is tempting to think that bodies are completely absent in this Marxist discourse, but, as the following example from Ron Johnston illustrates, bodies are present, albeit in less explicit ways than in the texts of the humanistic geographers.

Johnston (1984: 473) argues for a 'new' political geography to be informed by a 'realist philosophy of science, grounded in Marxist political economy'. He explains: 'The basic goal of all individuals is daily reproduction . . . At the most fundamental level, this means that individuals want societies which ensure sufficient food to maintain themselves and to produce viable children' (Johnston 1984: 474). Although the word 'body' is not used, Johnston is talking about bodies. He discusses bodies that need to eat and drink in order to live, bodies that have sex, women's bodies that give birth to the next generation. When the theoretical abstractions of Marxist geography are peeled back, it is possible to catch a glimpse of material, yet socially and politically encoded, bodies. Johnston refers to this level of the individual and collective body as 'the most fundamental level' and yet, like other Marxist geographers, he renders it as Other to that which is constructed as rational – the fleshless economy.

Many feminist geographers publishing in the 1980s also tended to render the body invisible and as Other to rationality and the economic realm. Feminist geographers tended to employ a dualism between sex (nature, biology, the body) and gender (nurture, socialisation) (see, for example, Foord and Gregson 1986; MacKenzie 1984, 1987; McDowell 1983 and Women and Geography Study Group of the IBG 1984). This distinction between sex and gender did not originate from feminist writings, rather, it was derived from the work of psychologist Robert Stoller (1968). It was, however, adopted by a number of influential feminist writers including Michele Barrett (1980), Nancy Chodorow (1978), Germaine Greer (1970), Kate Millett (1970) and Ann Oakley (1972). They used the distinction between sex and gender in order to argue that there are biological differences between the sexes at birth, but that it is primarily socialisation that results in women and men having different gender characteristics. The advantages of this conceptualisation were that it offered a way of distinguishing between the predetermined, innate characteristics of men and women and the other social differences. Further, it carried an evaluation of the social as the determinant of women's unequal position (see Gatens 1991b for a critique of this sex/gender distinction).

Louise Johnson (1990: 18) argues that there are also disadvantages to employing the sex/gender distinction in geography. One of these is 'the omission of the body as a vital element in the constitution of masculine and feminine identity and the consignment of those who argue for a "corporeal feminism" . . . into the nether world of biological essentialism'. Johnson (ibid) explains that geographers, 'in their zeal to avoid the accusation of biologism and by embracing the logics of historical materialism and liberalism, have ignored the possibilities of examining the sexed body in space'. Once again, the body functions as an Other in geographical discourse.

Obviously, within these broad subject areas there are likely to be large differences (including national, regional and individual differences) in the work carried out. But, generally speaking, the body has been both absent and present, denied and desired, repressed and possessed (see Wolff 1990: 121 cited in Rose 1993a: 32) in geographical discourses during the period from the 1960s to the 1980s. Different geographical discourses are unstable and have therefore both needed and rejected embodiment. They have been concerned with both sides of their constitutive oppositions (mind and body).

I have drawn examples from specific texts in order to build a case, but in doing this I do not mean to imply that the works I have chosen are any better or worse than numerous others published at similar times. Rather, I have attempted to choose texts that are reasonably typical of their genre. The body was not completely absent from disciplinary knowledges before the late 1980s/early 1990s but neither was it 'placed' centre-stage. The body has been an 'absent presence' (Shilling 1993). When examining examples of geographies typical of the 1960s, 1970s and 1980s it is possible to detect traces of the body (but not necessarily bodily traces).

Over the past decade a literature that foregrounds the politics of bodies and places has emerged in a number of disciplines, including geography. In the 1990s the body began to enjoy a new status. This does not mean that the process of Othering has suddenly stopped. A complex politics of knowledge was being articulated in the 1990s whereby the body is often present in geographical discourse but bears no messy traces of its materiality. I return to this point later in the chapter. Preceding this point, however, I examine some of the reasons for the body becoming so popular in the social sciences and humanities in the 1990s.

## Why now, this 'body craze'?

Kathy Davis (1997: 1) (drawing on 'a decade review' of literature on the body by Arthur Frank 1990) states that there have been different explanations put forward for this 'body craze' in the social sciences and humanities. Davis argues that some hold feminism responsible.

Since the 1970s feminists have generated an enormous amount of research on bodies, especially the female body.[8] The Boston Women's Health Book Collective published *Our Bodies, Ourselves* in 1971. This is an early example of the attention feminists have paid to bodies as sites of political contestation. Subjects such as rape, sexual violence, pregnancy, abortion and media images of women were high on the agenda of feminists working during the 1970s and 1980s. This earlier work was generated from a range of disciplines, theories and methodologies and focused on how (women's) bodies were implicated in a range of discursive and material practices (see Davis 1997: 4–15).

Davis (1997: 3 drawing on Frank 1990, 1991) argues that the current interest in embodiment has also been prompted by 'contradictions in academic discourse which have emerged in the wake of postmodernity'. She explains: 'In modernist discourse, the body represents the hard 'facts' of empirical reality, the ultimate justification for positivism and the Enlightenment quest for transcendental reason' (Davis 1997: 3–4). The body is the 'only constant in a rapidly changing world, the source of fundamental truths about who we are what is just and unjust, human and inhumane, progressive and retrogressive' (Frank 1990: 133).

As Davis (1997: 4) points out, however, the enormous variation in appearance, practices, comportment and motility of bodies in different cultures is also used by social scientists as a line of reasoning that bodies are socially constructed. Cultural variation in embodiment indicates the absurdity of the notion of a 'natural' body. Bodies, therefore, are 'an ideal starting point for a critique of universality, objectivity or moral absolutism' (ibid.). Bodies have also been used in postmodernist discourses. In postmodernist writings the material body is often replaced with a discursive or metaphorical body. Bodies tend to be used as sites from which to critique Enlightenment thinking and its tendency to treat reason and the masculine as the unmarked (and privileged) norm from which all else deviates: '. . . postmodern scholars, inspired by Foucault, take the body as the site *par excellence* for exploring the construction of different subjectivities or the myriad workings of disciplinary power' (ibid.). Both modernist and postmodernist writers use the body as a secure ground for claims about knowledge and truth. Frank (1991) argues that it is this use of the body for contradictory theoretical agenda that accounts for its current status in the social sciences and humanities.

Still other writers, such as Mike Featherstone (1983) and Bryan Turner (1984), argue that the current popularity of the body is due to radical changes in the cultural landscape of late modernity. Featherstone and Turner explain that while there has always been some attention placed on bodies, in late modernity the body has become the ultimate vehicle for writing one's identity. Bodies are means of self-expression – various meanings are attached to being fat, thin, muscular, blond, tattooed, pierced, wrinkled, tanned and so on. Davis (1997: 2 citing Crawford 1984: 80) points out that this interest in the body is coupled with recent medical advances and that 'Our bodies have become the ultimate cultural metaphor for controlling what is within our grasp' (see also Bordo 1993).

Regardless of the reasons for the upsurge of interest in the body, geographers are implicated in this broad trend. In some circles the body is now seen to be a trendy, 'cool', 'hip' area of study – one of the latest fashions in geography. In the section that follows I outline the contours of this new territory.

## Outlining the contours

I am certainly not the only geographer to remark on the upsurge of interest in the body. Rose (1995: 545) claims that there is 'a growing concern with the bodily' in geography and that 'an interest in the corporeal is becoming evident in a range of studies'. Felicity Callard (1998: 387) claims: 'Geographers are now taking the problematic of corporeality seriously. "The body" is becoming a preoccupation in the geographical literature, and is a central figure around which to base political demands, social analyses, and theoretical investigations.' There has been a 'dash towards things corporeal' (Callard 1998: 388).

Michael Dorn and Glenda Laws (1994: 109) argue that feminists and cultural theorists have 'confronted the politicization of bodies in their work and slowly geographers are beginning to take up the challenge'. Laws (1997: 49; emphasis in original) explains that: 'Geographers need to attend to both the conceptualisation and material construction of bodies because *our bodies make a difference to our experience of places:* whether we are young or old, able-bodied or disabled, Black or White in appearance does, at least partly, determine collective responses to our bodies'. Mona Domosh (1997: 85) comments that: 'emphasis on understanding the politics of places through personal stories, making sense of our embodied persons and spaces through the political, seems key to a more "clear" geography'. Domosh makes this claim in relation to feminist geography but I think her comments also apply more generally across the discipline.

Several feminist geographers have posed interesting questions about the potential of examining the relationship not just between bodies and spaces but between gendered/sexed bodies and spaces. Linda McDowell (1993: 306) urges geographers to consider the body more carefully and suggests that 'the implications of these differences [between men's and women's bodies] for geographical concepts of spatiality, boundaries and community remain to be explored'. In *Gender, Identity and Place: Understanding Feminist Geographies* (1999) McDowell devotes a chapter to 'In and out of place: bodies and embodiment'. J. K. Gibson-Graham (1996), prompted by Sharon Marcus's (1992) work on rape, poses questions about the parallels between women's bodies being scripted as 'lack' (awaiting rape) and conceptualisations of global capitalism as penetrating. Gibson-Graham (1996: 124) claims: 'There are many obvious points of connection between the language of rape and the language of capitalist globalization'. She prompts readers to think about how new geographies might be created by placing these discourses side by side. Johnson (1994a: 107) asks: what might happen to our geographies if we return to the work of humanistic geographers in the 1970s (who drew on theories of phenomenology) and add this to 'feminist concern[s] for sexualised bodies?'

There has also been some work by geographers on the 'racialisation' of bodies. Much of this focuses on 'black' or 'coloured' bodies rather than on 'white' bodies. For example, Jackson (1994: 57) discusses some of the ways in which the image of black male bodies played a crucial role in repositioning 'Lucozade' (a brand name soft drink). Elder (1995) examines the codification of sexualised/racialised bodies in apartheid South Africa. Jarosz (1992) looks at metaphors employed to construct Africa as the dark continent. Radcliffe (1999) examines the connections between racialised subjects, gender and national identity in contemporary Ecuador. Her focus rests on *mestizo* men and white women.

Over the last decade there has also been an increasing amount of research carried out on the sexual body. In 1992 David Bell established a network called 'Sexuality and Space'. Also in 1992 Beatriz Colomina edited a collection of essays published in a volume entitled *Sexuality and Space*. In 1999 the *Journal of Geography in Higher Education* published a symposium on teaching sexualities in geography. Research by people such as Sy Adler and Johanna Brenner (1992), David Bell (1991), David Bell and Gill Valentine (and their contributors) (1995), Bell *et al.* (1994), Jon Binnie (1997), Stephen Hodge (1995), Lawrence Knopp (1990a, 1990b, 1992), Ki Namaste (1996) and Gill Valentine (1993, 1997) are playing a vital role in retheorising geography – a retheorising that involves problematising the mind/body dualism and making bodies (sexual bodies) explicit in the production of geographical knowledge.

At conferences such as those of the Association of American Geographers (AAG), the Institute of Australian Geographers (IAG), the New Zealand Geographical Society (NZGS) and the Royal Geographical Society/Institute of British Geographers (RGS/IBG) individual papers and whole sessions are now devoted to the body. As early as 1989 Johnson presented a paper to the New Zealand Geographical Society entitled 'Embodying geography – some implications of considering the sexed body in space'.

Also, over the last few years, geographers have produced a number of books that focus on corporeality. Examples of authored books include *Sensuous Geographies: Body, Sense and Place* by Paul Rodaway (1994), *The Body and the City: Psychoanalysis, Space and Subjectivity* by Steve Pile (1996), and *Consuming Geographies: We are Where we Eat* by David Bell and Gill Valentine (1997). Examples of edited books include: *Mapping Desire: Geographies of Sexualities* by David Bell and Gill Valentine (1995), *Body Space: Destabilising Geographies of Gender and Sexuality* by Nancy Duncan (1996), *Places Through the Body* by Heidi Nast and Steve Pile (1998), *New Frontiers of Space, Bodies and Gender* (1998) by Rosa Ainley, *Embodied Geographies* (1999) by Elizabeth Teather, and *Mind and Body Spaces* (1999) by Ruth Butler and Hester Parr.

There are also a number of geography books that do not necessarily mention the word 'body' in their titles but address issues of embodiment in parts of the text. For example, in *Cool Places: Geographies of Youth Cultures* (edited by Tracey Skelton and Gill Valentine), Birgit Richard and Heinz Hermann Kruger (1998: 169) discuss 'new masculinities on the dance floor' in Berlin; Ben Malbon (1998: 270) examines clubbing as 'a sensational performance'; Ruth Butler (1998: 83–100) rehabilitates images of disabled youths. Linda McDowell turns over the second half of her book *Capital Culture: Gender at Work in the City* (1997) to 'Bodies at work'. Catherine Nash discusses 'Remapping the body/land: new cartographies of identity, gender, and landscape in Ireland' in an edited collection by Alison Blunt and Gillian Rose (1994: 227–250). In *The End of Capitalism (as we knew it)* J. K. Gibson-Graham (1996: 135) suggests rewriting the male body and sexuality in ways that 'might affect views of capitalism and its globalizing capacities' (see chapter 6, 120–147). Many of the authors, including Glenda Laws, Richa Nagar, Laura Pulido and Lydia Mihelic Pulsipher, in *Thresholds in Feminist Geography* (edited by Jones, Nast and Roberts 1997) make the point that bodily experiences are spatially lived.

Numerous articles on the body have also appeared in geographical journals. In *Gender, Place and Culture* David Bell, Jon Binnie, Julia Cream and Gill Valentine (1994: 31) 'think about the performance of sexual identities in space' focusing on the 'hypermasculine "gay skinhead" and the hyperfeminine "lipstick lesbian"'. In *Society and Space* Linda McDowell and Gill Court (1994) write about 'Performing work: bodily representations in merchant banks'. Also in *Society and Space* Jon

Binnie (1997) discusses 'Coming out of geography: towards a queer epistemology?' In *Progress in Human Geography* Gillian Rose (1995) reports on 'Geography and gender: cartographies and corporealities'. In the *New Zealand Geographer* Lynda Johnston (1997) examines 'Queen(s') Street or Ponsonby Poofters? Embodied HERO parade sites'.

This research is also finding its way into university geography teaching. In the Department of Geography at the University of Waikato I teach a graduate course entitled 'Crossing boundaries'. The course reflects geographers' burgeoning interest in the body as a social phenomenon. Michael Brown in the Department of Geography offers a graduate course on 'Geographies of the body' at the University of Washington, Canada. In the Department of Geography, University of Edinburgh, in 1997/98, Gillian Rose states one of the aims of a course entitled 'The cultural politics of landscape' is to examine some of the 'ways in which visual spaces of place and embodiment are produced through power relations' (course handbook 1997–8: 1). In the same department Liz Bondi and Lynda Johnston offer an 'honours option' course entitled 'Gender, sexuality and the city' in which students are given an opportunity to examine 'embodied experiences of the city'.

This list of courses and publications that focus on bodies and spaces is certainly not exhaustive. Nor is it meant as a review. I have chosen examples to indicate that geographers are interested in the body. There is also evidence of a 'body craze' in a range of other disciplines including feminist studies, cultural studies, postcolonial studies, health studies and sociology. Some writers working in these subject areas pay attention to spatiality. For example, art historian Sue Best (1995) in a chapter entitled 'Sexualising space' examines the ways in which space is conceived as a woman. Theorists, such as Moira Gatens (especially 1988, 1991a, 1991b) and Elizabeth Grosz (especially 1992), have written extensively about embodiment and have paid careful attention to space. Embodiment and space is an interdisciplinary problematic which is attracting the scrutiny of geographers as well as many others.

## Taking flight from fluids and mess

Like Callard (1998) I am interested in the manner in which some ways of theorising and understanding the corporeal in geography in the 1990s have gained predominance. Callard (1998: 388) elaborates:

> . . . how the body acts as a methodology by summoning up certain theoretical imperatives in the very mention of corporeality. The call to understand the importance of the body is often simultaneously a call for the fluidity of subjectivity, for the instability of the binary of sexual difference, and for a host of other working assumptions.

Callard argues that 'the body' or 'mention of corporeality' in geography has become a shorthand for a number of theoretical imperatives, one of which is 'fluidity of subjectivity'. I agree with Callard but want to highlight a particular irony here. Although the body is used to refer to fluid subjectivity (and identity), geographers seldom refer to the actual materiality and fluidity of the body itself. While it has become highly acceptable to employ postmodernist metaphors of fluidity and mobility, it is still not acceptable for the flesh and boundaries of fluid, volatile, messy, leaky bodies to be included in geographical discourse.

Numerous geographers in the 1990s who were influenced by postmodern theorists such as Judith Butler (1990, 1993), Gilles Deleuze and Felix Guattari (1983, 1986), and Michel Foucault (1979, 1980) now conceive of identity as fluid. The fluid, volatile flesh of bodies, however, tends not to be discussed. There is little in the discipline that attests to the runny, gaseous, flowing, watery nature of bodies. The messy surfaces/depths of bodies, their insecure boundaries, the fluids that seep and leak from them, that which they engulf, the insides and outsides that sometimes collapse into each other remain invisible in the geographical canon. When geographers speak of the body they still often fail to talk about a body that breaks its boundaries – urinates, bleeds, vomits, farts, engulfs tampons, objects of sexual desire, ejaculates and gives birth. The reason this is significant is that the messiness of bodies is often conceptualised as feminised and as such is Othered. Bracketing out questions about the boundaries of body/space relationships functions as an attempt to position geographical knowledge as that which can be separated out from corporeality, the corporeality of its subjects and its producers. Ignoring the messy body is not a harmless omission, rather, it contains a political imperative that helps keep masculinism intact.

This exclusion of the material body may be, in part, both a reflection and result of social constructionism which has gained recognition in the discipline over the last decade. Social constructionists sometimes depict bodies as though they were little more than surfaces etched with social messages. Having said this, however, in other ways social constructionism has offered a great deal to geographers. Social constructionism has helped destabilise the longstanding notion that bodies are 'simply natural' or biological. It has also reiterated the point that bodies cannot be understood outside of *place* (see Grosz 1992). One of the downsides of social constructionism though is that is can render the body incorporeal, fleshless, fluidless, little more than a linguistic territory. The materiality of bodies becomes reduced to systems of signification.

I am not alone in making this point about the exclusion of materiality in recent geographical work on the body. Dorn (1998: 183; emphasis in original) drawing on an argument made by Bordo (1993) explains: 'the proliferation of postmodernist metaphors of nomadism – hybrids, freaks, cyborgs, monsters, tricksters and *mestizas* – signals a dangerous retreat from an earlier emphasis on *embodied* located critiques.' By 'earlier emphasis on embodied located critiques' I suspect that Dorn is referring to phenomenological accounts of the lived body prevalent in humanistic geographies in the 1970s. Dorn argues that feminist poststructuralist work on shifting locations, fluid and multiple identities and nomadic forms of thought is underpinned by ableist assumptions. He urges geographers to keep in sight the material environments that 'real' bodies must negotiate. We must not take 'flight from the messiness of disability into myth and metaphor' (Dorn 1998: 184). Of course it is not only disabled bodies, but *all* bodies, that are messy (although some are considered messier than others). Dorn correctly points out that disabled persons are likely to find numerous material environments more difficult to negotiate than able-bodied persons, but every body has a weighty materiality and boundaries that are enmeshed in specific social and cultural systems of signification. There are no neat binary divisions between disabled and able-bodied persons any more than there are between mind and body or between the inside and outside of bodies.

While social constructionist approaches tend to focus on the exteriorities of bodies, psychoanalytic approaches tend to articulate bodies as internal and psychical spaces.[9] This is a rather sweeping statement since psychoanalytic theory is clearly an immense and broad field. Different authors have drawn on different theorists and approaches. Pile (1996), for example, uses object-relations theory, theories of abjection and Lacan's concept of the Other in *The Body and the City*. Liz Bondi (1999) opens up a useful discussion about the interconnections between psychotherapeutic practice and human geography. Despite these huge differences in approach and the relative merits of each, the fact remains that geographers employing these approaches have said little about the seemingly banal bodily acts such as farting, shaving and urinating and the significance of understanding these in terms of body/politics/space relationships. Although psychoanalytic approaches have helped many of us understand concepts such as 'lack' in relation to sexual difference and some discussions have focused on the insecurity of boundaries (Pile 1998 destabilises the boundary between the conscious self and the dynamic unconscious by discussing dreams), the potential of these approaches has not yet really been tapped by geographers.

Humanist geographers using phenomenological approaches have also failed to address the points at which bodies' interiorities and exteriorities, depths and surfaces come together or apart. Humanist geographers examine people's emotions and 'life worlds' but not the leakiness of bodily boundaries, the constructedness of dirt or bodily mess. Many topics remain taboo for humanist geographers, for example, the life worlds of lactating women, vomit on the streets of inner cities or what people do in domestic bathrooms. Paul Rodaway (1994) calls upon theories of postmodernism and humanism (phenomenology) in his book *Sensuous Geographies* and yet the bodies invoked in the text seem to lack substance. Rodaway (1994: 4) states that the primary aim of his book is to 'excite interest in the immediate sensuous experience of the world and to investigate the role of the senses – touch, smell, hearing and sight – in geographical experience'. He focuses on 'perception theory' and examines specific sensory geographies – haptic, olfactory, auditory and visual. Rodaway also discusses some of the ways in which sensuous experiences of place have been transformed recently by various technologies and cultural practices. Although there is a great deal of interesting material contained in the book, my reading of it left me disconcerted by a feeling that the bodies discussed seemed more like cardboard cut-outs rather than flesh and blood bodies that fart, weep and make love. The text and diagrams used throughout seemed to 'tidy up' much of the messiness and irrational sensuousness of human experience and corporeality (Longhurst 1994b).

Ignoring the body that has volatile boundaries, regardless of the approach adopted, carries several 'costs'. The first of these costs is that many themes and topics remain invisible and/or are deemed illegitimate by the hegemons in the discipline. Topics such as the geographies of 'the homeless' (including sores and crumbling teeth) (although see Peace 1999), or geographies of domestic spaces such as bedrooms (including sleep and romance) are deemed to be overly subjective and 'non-academic'. They threaten to spill, soil and mess up, clean, hard, masculinist geography. Geography's boundaries undergo constant securing in an attempt to legitimise particular (disembodied) knowledges. In this process a great deal is edited out – considered to be dirty or just plain banal and inappropriate. Preference is given to the clean, clinical, important, scientific, statistical and heroic. What constitutes appropriate issues and legitimate topics to teach and research in geography comes to be defined in terms of reason, rationality and transcendent visions, as though these can be separated out from passion, irrationality, messiness and embodied sensation.

The cost of geography shunning dirty topics/messy bodies is borne by those people who desire to examine such topics. To date, many of the themes, topics and approaches that have been adopted in geography have been those that address the needs and interests of men, in particular, white, bourgeois, able-bodied men. This is not surprising because as Rose (1993a: 1) notes: 'The academic discipline of geography has historically been dominated by men, perhaps more so than any other science'. People who want to address dirty (Other) topics, people who themselves may be defined as Others (such as the ill, frail, diseased, homosexual, elderly, black, poor, disabled, working class – bodies that are often thought to be messy and out of control), are forced to struggle for legitimisation of their interests in the discipline.

A defensive line between inside and outside the discipline is continually being (re)drawn. This is an issue not just about what counts as geographical knowledge but also about *who* counts as a bearer of geographical knowledge. Although we all have bodies only those people who conceptually occupy the place of the mind are 'thought' to be able to produce such knowledge. For those people who are constructed by Cartesian philosophy as being tied to their bodies, transcendent visions are not considered possible. Their knowledge cannot count as knowledge for it is too intimately grounded in, and tainted by, their corporeality. Those people whose bodies are understood to be messy and/or out of control – the disabled, pregnant, lactating, dirty, queer, fat, elderly, poor – are likely to be marginalised as illegitimate bearers of geographical knowledge.

Rose (1993a: 4) argues that it is possible that there is something in the very claim to knowing in geography that tends to exclude women as producers of knowledge. There are numerous others who could also be put into the category of 'excluded' (see Sibley 1995). As geographers attempting to understand something about place it is significant that each of us has skin that is a certain colour, genitalia that are sexually specific, a sexual orientation of sorts, bodies with specific (in)capacities, and yet despite (feminist/postmodernist) calls to 'situate' our knowledges we still rarely do acknowledge (taint?) our knoweldges with messy traces of our corporeality. At conferences I sometimes attempt to match up the flesh of authors with their textual productions. Often the bodies come as a surprise, maybe because 'The knowing subject who produces knowledge . . . [has been] bracketed off from the knowledges thus produced' (Grosz 1993: 191).

Although it now permissible, even desirable, within feminist and postmodernist discourse to acknowledge something of ourselves and our political locatedness in the production of our texts (as complex as this is – see Rose 1997) it is still permissible only to acknowledge and reflect on certain things about ourselves – things we are supposed to unproblematically know and understand such as our gender, ethnicity, and/or age. Many things remain off limits – too private, too 'inappropriate', too messy, to put into our epistemic 'master' pieces.

Three separate narratives spring to 'mind' as ways of situating myself. The conventional narrative offers my ethnicity, gender and age, thus I construct a story that I am Pakeha,[10] female, and 38. A less conventional narrative, and one that might now be more noticed, is my size, food preference and bodily habits, thus I could situate myself as an 'over' weight 'chocoholic' who refuses to exercise. Or again, I might choose my parental status to present a narrative about my youngest son's refusal to use the toilet and my role as a frustrated parent.

Both these later stories of my body shape and my relationship with my son are significant to the way in which I understand (public and private) places and both play a significant role in the writing of this geography. The point that I want to stress here though, is that neither of these things (my body or my son's body) are commonly voiced in geographical texts or forums.

It is not that they are simply too private. Anyone who sees me knows that I am 'large' and almost anyone who knows me has heard me complain about my son's refusal to use the toilet. Some might argue that these things belong in the category of the everyday, prosaic and banal. Perhaps they do. But banal does not necessarily imply boring (that is, intellectually uninteresting and unimportant). The banal ought not to escape attention or be sidelined as domestic, feminine and Other. In discussing my body and my son's body I begin to establish my relationship with Others. I situate myself, not as an autonomous, rational academic but as an (irrational) mother who is intimately connected in every way (emotionally, bodily) to my son. The type of positioning that I have engaged in also conveys an identity that is shifting. It is possible that I will not always be large and (hopefully) soon I will not be someone who has to clean up my son's excrement every day. Writing myself as 'large' or 'over' weight means acknowledging the volatility of my bodily boundaries as I continually gain/lose weight. Large people are often seen as 'out of control'. My son's body might also be understood to be 'out of control'. This kind of situating or positioning resists the god-trick. I am talking from an embodied place rather than from a place on high. I am positioned to talk about not just that which is rational and public but also that which is irrational and private.

I do not want to give the impression that *all* geographers who have focused attention on bodies have treated them as mess- and matter-free. I have already mentioned Dorn's (1998) work on 'crippled' bodies. There are also a number of other authors working in the area of geographies of illness, impairment and disability who acknowledge the weighty and messy materiality of bodies. For example, Vera Chouinard and Ali Grant (1996) write their own material bodies into 'the project'. Isobel Dyck (1999) highlights the 'body troubles' experienced by women with chronic illness, specifically MS, *multiple sclerosis*, in their workplaces (see also Moss and Dyck 1999 on women diagnosed with ME, *myalgic encephalomyelitis*, popularly referred to as Chronic Fatigue Syndrome).

David Bell and Gill Valentine (1997) examine the ways in which eating conveys embodied identity. A fascinating (and, for a geography book, rather unusual) illustration graces the cover of Bell and Valentine's book *Consuming Geographies*. A fat man delights in licking cream from the stubbled skin around the edges of his lips and mouth. He is depicted as enjoying the sensual pleasure of eating. His eyes are closed as he relishes the last morsels of cream. However, the slightly distorted image also depicts him as grotesque and excessive. His large body breaks it boundaries (his tongue protrudes) in a display of desire and repulsion.

David Sibley (1995) also discusses the breaking of bodily boundaries and the constructions of some groups, such as Jews and Gypsies, as abject Others. Tim Cresswell (1997) analyses media representations of women at the Greenham Common women's peace camp. He notes that the media report the camp as being surrounded by sanitary napkins. Cresswell (1999) also discusses the bodies of 'female tramps and hobos'.

In the remainder of this book I build on these aforementioned geographies to discuss the volatile materiality of bodies and spaces. Notions of abjection, boundaries, fluids and dirt prove useful for this purpose. In the following section I outline work by Anne McClintock (1995), Elizabeth Grosz (1994a), Iris Young (1990a), Mary Douglas (1966) and Julia Kristeva (1982). Their ideas provide a basis for the following chapters on pregnant women in public spaces, heterosexual, 'white' men in domestic bathrooms and managers in CBDs.

## Abjection

Following Anne McClintock's (1995: 71–74) lead in her excellent book *Imperial Leather*, I develop a 'situated psychoanalysis' paying specific attention to the notion of abjection. McClintock (1995: 73) convincingly argues that '. . . psychoanalysis and material history are mutually necessary for a strategic engagement with unstable power'. She explains that psychoanalysis needs to be culturally contextualised and informed by history. Psychoanalytic theory has been criticised for accepting universalist assumptions that identity formation is essentially human rather than culturally, spatially and temporally specific. Situating it would help avoid this pitfall. McClintock also argues that history ought to be informed by psychoanalysis. She explains that: 'Abjection shadows the no-go zone between psychoanalysis and material history, but in such a way as to throw their historical separation radically into question' (McClintock 1995: 72).

> Abjection (Latin, *ab-jicere*) means to expel, to cast out or away. In *Totem and Taboo* and *Civilizations and its Discontents* Freud was the first to suggest that civilization is founded on the repudiation of certain pre-oedipal pleasure and incestuous attachments.
>
> (McClintock 1995: 71)

Kristeva also examines the notion of abjection. In her book *Powers of Horror* (1982) she studies numerous personalised bodily horrors. These horrors mark the significance for subjects (subjects as they exist within certain cultures) of the various boundaries and orifices of the body. Kristeva questions the conditions under which the proper, clean, decent, obedient, law-abiding body is demarcated and emerges. The cost of the clean and proper body emerging is what Kristeva terms abjection. Abjection is the affect or feeling of anxiety, loathing and disgust that the subject has in encountering certain matter, images and fantasies – the horrible – to which it can respond only with aversion, nausea and distraction. Kristeva argues that the abject provokes fear and disgust because it exposes the border between self and other. This border is fragile. The abject threatens to dissolve the subject by dissolving the border. The abject is also fascinating, however; it is as though it draws in the subject in order to repel it (see Young 1990a: 145).

Grosz (1994a: 192), in discussing Kristeva's work on abjection, claims:

> The abject is what of the body falls away from it while remaining irreducible to the subject/object and inside/outside oppositions. The abject necessarily partakes of both polarized terms but cannot be clearly identified with either.

The abject is undecidable, both inside and outside. Kristeva uses the example of 'disgust at the skin of milk' (Grosz 1989: 74) – a skin which represents the subject's own skin and the boundary between it and the environment. Abjection signals the tenuous grasp 'the subject has over its identity and bodily boundaries, the ever-present possibility of sliding back into the corporeal abyss out of which it was formed' (Wright 1992: 198). In ingesting objects into itself or expelling objects from itself, the subject can never be distinct from the objects. These ingested/expelled objects are neither part of the body nor separate from it. The abject (including tears, saliva, faeces, urine, vomit, mucus) marks bodily sites/sights which will later 'become erotogenic zones' (mouth, eyes, anus, nose, genitals) (Grosz 1989: 72; see also Wright 1992: 198).

McClintock (1995: 72; emphasis in original) suggests:

> With respect to abjection, distinctions can be made, for example, between abject *objects* (the clitoris, domestic dirt, menstrual blood) and abject *states* (bulimia, the masturbatory imagination, hysteria), which are not the same as abject *zones* (the Israeli Occupied Territories, prisons, battered women's shelters). Socially appointed *agents* of abjection (soldiers, domestic workers, nurses) are not the same as socially abjected *groups* (prostitutes, Palestinians, lesbians). *Psychic* processes of abjection (fetishism, disavowal, the uncanny) are not the same as *political* processes of abjection (ethnic genocide, mass removals, prostitute 'clean ups').

These are distinct dimensions, but also interdependent, elements of abjection. They are not transhistorical and universal but, rather, are interrelated and, in some instances, contradictory elements of a complex process of psychic and social formation.

Young (1990a: 142) makes effective use of the category 'socially abjected groups' to argue that some groups are constructed as 'ugly'. Young (1990a: 145) argues that understanding abjection enhances 'an understanding of a body aesthetic that defines some groups as ugly or fearsome and produces aversive reactions in relation to members of those groups'. Young (ibid.) states: 'Racism, sexism, homophobia, ageism and ableism are partly structured by abjection, an involuntary, unconscious judgement of ugliness and loathing'.

A great deal of the work on abjection is anchored by Douglas's insights on boundary rituals and dirt. Douglas (1975: 47–59) argues that nothing in itself is dirty, rather, dirt is that which is not in its proper place and upsets order. Douglas (1966: 5) claims: 'Reflection on dirt involves reflection on the relation of order to disorder, being to non-being, form to formlessness, life to death'. Dirt is essentially disorder – it is 'matter out of place'.

> If we can abstract pathogenicity and hygiene from our notion of dirt, we are left with the old definition of *matter out of place* . . . Dirt then, is never a unique, isolated event. Where there is dirt there is system. Dirt is the by-product of a systematic ordering and classification of matter, in so far as ordering involves rejecting inappropriate elements. This idea of dirt takes us straight into the field of symbolism and promises a link-up with more obviously symbolic systems of purity.
>
> (Douglas 1966: 35; my emphasis)

Grosz (1994a: 192 and 202) uses Douglas's ideas on dirt and Kristeva's notion of abjection in order to explore the 'powers and dangers' of body fluids.

In the following paragraph Grosz succeeds in capturing something of the disquiet about and unsettling nature of body fluids or corporeal flows – tears, amniotic fluids, sweat, pus, menstrual blood, vomit, saliva, phlegm, seminal fluids, urine, blood. For this reason I quote her at length.

> Body fluids attest to the permeability of the body, its necessary dependence on an outside, its liability to collapse into this outside (this is what the death implies), to the perilous divisions between the body's inside and outside. They affront a subject's aspiration toward autonomy and self-identity. They attest to a certain irreducible 'dirt' or disgust, a horror of the unknown or the unspecifiable that permeates, lurks, lingers, and at times leaks out of the body, a testimony of the fraudulence or impossibility of the 'clean' and 'proper'. They resist the determination that marks solids, for they are without any shape or form of their own. They are engulfing, difficult to be rid of; any separation from them is not a matter of certainty, as it may be in the case of solids. Body fluids flow, they seep, they infiltrate; their control is a matter of vigilance, never guaranteed.
>
> (Grosz 1994a: 193–194)

Fluids are 'enduring'; they are 'necessary' but often 'embarrassing' within western cultures – they are frequently considered to be undignified 'daily attributes of existence' that we all must, although in different ways, live with and reconcile ourselves to (Grosz 1994a: 194). The fluids that cross bodily boundaries between inside and outside include tears, saliva, faeces, urine, vomit, sweat and mucus. These fluids often provoke feelings of abjection.

But bodily fluids are not all the same. Grosz (1994a: 195) notes that they have 'different indices of control, disgust and revulsion. There is a kind of hierarchy of propriety governing these fluids themselves.' Some 'function with clarity', that is, they are 'unclouded by the spectre of infection' and 'can be represented as cleansing and purifying' (ibid.). For example, tears do not carry with them the 'disgust associated with the cloudiness of pus, the chunkiness of vomit, the stickiness of menstrual blood' (ibid.). The latter are seen as polluting fluids that mess up the body whereas clean fluids, such as tears, are often considered to cleanse the body (see also Douglas 1966: 125). Although there may be bacterial properties associated with specific body fluids – the 'real' body and the micro-organisms it houses cannot be denied – there is not *necessarily* anything inherently polluting or cleansing about specific body fluids.

Douglas (1966: 38) refers to Sartre's analysis of the viscous in *Being and Nothingness* as a part explanation of 'our' horror of bodily fluids. Grosz (1994a: 194) claims that: 'For both Douglas and Sartre, the viscous, the fluid, the flows which infiltrate and seep, are horrifying in themselves'. Douglas quotes from Sartre's essay on stickiness (1956), in which he argued that viscosity repels in its own right as a primary experience. Sartre (1956 cited in Grosz 1994a: 194) explains that: 'The viscous is a state half-way between solid and liquid. It is like a cross-section in a process of change . . . to touch stickiness is to risk diluting myself into viscosity. Stickiness is clinging, like a too possessive dog or mistress.'

Grosz (1994a: 194) points out that: 'Like Sartre, Douglas associates this clinging viscosity with the horror of femininity, the voraciousness and indeterminacy of the *vagina dentata*'. It is evident that 'this fear of being absorbed into something which has no boundaries of its own, is not a property of the viscous itself' (Grosz 1994a: 194). Like dirt, the viscous and the fluid refuse to conform to the laws governing the proper, the clean and the solid. The viscous is liquid/matter that will not stay in place. Female sexuality is not inherently or essentially viscous, rather, 'it is the production of an order that renders female sexuality and corporeality marginal, indeterminate, and viscous that constitutes the sticky and the viscous with their disgusting, horrifying connotations' (Grosz 1994a: 195).

Irigaray (1985) argues that this unease about viscosity is linked to the fact that it is not possible to speak of indeterminacy, ambiguity and fluidity within prevailing western philosophical models of being. Fluids are implicitly associated with femininity, maternity, pregnancy, menstruation and the body. Fluids are subordinated to that which is concrete and solid. In turn, solidity and rationality become linked (Irigaray 1985: 113). 'Douglas refers to all borderline states, functions, and positions as dangers, sites of possible pollution or contamination' (Grosz 1994a, 195). Douglas conceptualises fluid as a borderline state, as liminal, and as disruptive of the solidity of things and objects (ibid.).

Clearly, bodies and their associated fluids are not simply natural or given but rather represent social relations. Their orifices and surfaces symbolise 'sites of cultural marginality, places of social entry and exit, regions of confrontation or compromise' (Grosz 1994a: 193). Lived experiences of body fluids are mediated through cultural representations and through sex/gender.

Grosz aims to further understanding of notions of sexual difference and the culturally coded meanings of body fluids and viscosity. She argues that body fluids, and in particular 'women's corporeal flows' (Grosz 1994a: 202), act as markers of sexual difference. Douglas (1966: 3–4 cited in Grosz 1994a: 193) argues that 'there are beliefs that each sex is a danger to the other through contact with sexual fluids' yet only one sex tends to be 'endangered by contact with the other, usually males from females'. So far, in most western cultures, men's body fluids have not been regarded as polluting and contaminating for women in the same way as women's have been for men. Grosz (1994a: 197) notes: 'It is women and what men consider to be their inherent capacity for contagion, their draining, demanding bodily processes that have figured so strongly in cultural representations, and that have emerged so clearly as a problem for social control'. Grosz (1994a, 202) argues that 'only when men take responsibility for and pleasure in the forms of seepage that are their own, when they cease to reduce it to its products [for example reducing the formless fluidity of semen to the solid form of a fetus – *his* property]. When they accept their sexual specificity' will they be able to respect women's bodily autonomy and sexual specificity.

In the chapters that follow I draw on the notion of abjection. I do not, however, discuss and people and places commonly regarded as abject. (McClintock 1995: 72 mentions people such as slaves, prostitutes, the colonised, domestic workers, the insane, the unemployed and places such as slums, ghettos, garrets and brothels.) Rather, I discuss people and places often regarded as respectable. First, I focus on pregnant women and their relationship with public space (Chapter 3). Second, I focus on heterosexual, able-bodied, Pakeha men and their experiences of domestic toilets/bathrooms (Chapter 4). Third, I focus on managers working in CBDs (Chapter 5). These chapters illustrate that abjection functions, albeit in different ways, in relation to all bodies and spaces, not just those regarded as Other.

# 3

# PREGNANT BODIES IN PUBLIC PLACES

Pregnant women in public space are often constructed as 'matter out of place' (Douglas 1966: 35). In this chapter I draw on the earlier discussion of abjection to further understanding of a group of women who were pregnant for the first time and living in Hamilton, New Zealand (see Longhurst 1995a, 1998). Pregnant women are thought to threaten and disrupt a social system that requires them to remain largely confined to private space during pregnancy. They can be seen to occupy a borderline state as they disturb identity, system and order by not respecting borders, positions and rules. Some work which links pregnancy and abjection has already been carried out. Oliver (1993), for example, in *Reading Kristeva* discusses 'the abject mother'. Jan Pilgrim (1993) uses Kristeva's concept of the abject to examine representations of the naked pregnant body. I utilise Young's notion of 'ugly bodies' by examining the possibility that pregnant bodies are sometimes constructed, both by pregnant women themselves and by others, as ugly.

Over a period of approximately two years – May 1992 through to July 1994 – I talked with 31 women who were pregnant for the first time and living in Hamilton, New Zealand (see Table 3.1). Hamilton is a city of 132,104 people (Census of Population and Dwellings 1996) – the fourth-largest city in the country. It is located to the west in the northern half of the North Island of New Zealand. Hamilton is the major commercial and industrial centre servicing surrounding agricultural and pastoral land. The Waikato region has a population of approximately 350,000. The city and the outlying regions are serviced by the Waikato Women's Hospital where 3,273 women gave birth in 1993. A very small proportion – estimated at between 2 and 5 per cent – of babies are born at home. Most of the buildings in Hamilton's CBD are no more than six or seven storeys high, and roads and footpaths are reasonably wide and uncrowded (see Plate 3.1). This spatial form means that many people (but certainly not all, including pregnant women) find the city to be reasonably accessible. More than 80 per cent of 'Hamiltonians' are of Anglo-European descent, 11 per cent are Maori, the remaining 9 per cent is made up of a diverse range of many cultures including Asian, African and American.

I talked generally with women about their experiences of pregnancy. I asked

*Table 3.1* Pregnant women living in Hamilton, New Zealand, 1992–1994: profile of participants' general characteristics.

| | *Age* | *Ethnicity* | *Weeks pregnant* | *Occupation* |
|---|---|---|---|---|
| *In-depth case-study interviews* | | | | |
| Denise | 25–29 | Pakeha | 8+ | Domestic worker |
| Kerry | 20–24 | Pakeha | 15+ | Training consultant |
| Paula (& husband Roy) | 20–24 | Pakeha | 15+ | Full time at home |
| Sarah | 15–19 | Maori | 15+ | Enrolled in work training courses |
| *One-off individual interviews* | | | | |
| Jude | 25–29 | Pakeha | 33 | Tertiary student |
| Katie | 30–34 | Pakeha | 34 | Self employed – clothing business |
| Michelle | 30–34 | Pakeha | 33 | Dance teacher |
| Mary Anne | 25–29 | Pakeha | 36 | Travel consultant |
| Sonya | 25–29 | Pakeha | 26 | Singer |
| Dorothy | 25–29 | Pakeha | 40 | Sales representative |
| Ngahuia | 25–29 | Maori | 39 | University lecturer |
| Sandy | 15–19 | Pakeha | 26 | Checkout operator |
| Christine (& husband Howard) | 25–29 | Pakeha | 38 | Bank teller |
| Helen (& husband Gary) | 20–24 | Maori | 31 | Tertiary student |
| Tracy (& partner Dan) | 25–29 | Pakeha | 39 | Secretary |
| *Focus group 1* | | | | |
| Donna | 15–19 | Pakeha | 29 | Receptionist |
| Sam | 30–34 | Pakeha | 36 | Full time at home |
| Jill | 25–29 | Pakeha | 30 | Office worker |
| Stella | 15–19 | Pakeha | 36 | Book store worker |
| Penny | 30–34 | Pakeha | 29 | Government dept. worker |
| *Focus group 2* | | | | |
| Adrienne | 20–24 | Pakeha | 39 | Full time at home |
| Joan | 25–29 | Pakeha | 41 | Nurse |
| Moira | 25–29 | Pakeha | 34 | School dental nurse |
| Sharon | 30–34 | Pakeha | 31 | Full time at home |
| Joanne | 20–24 | Pakeha | 38 | Receptionist |
| *Focus group 3* | | | | |
| Margaret | 25–29 | Pakeha | 39 | Childcare worker |
| Terry | 30–34 | Pakeha | 25 | Tertiary student |
| *Focus group 4* | | | | |
| Dawn | 15–19 | Pakeha | 29 | Machinist |
| Angela | 20–24 | Pakeha | 35 | Full time at home |
| *Focus group 5* | | | | |
| Moana | 20–24 | Maori | 32 | Personal assistant |
| Rebecca | over 34 | Pakeha | 32 | Risk management assessor |

*Notes*
+ Denotes 'until full term'.
This profile excludes the five participants who were involved in preliminary interviews.

*Plate 3.1* Hamilton CDB – photomontage of areas frequented by participants.
*Source:* Photographs by Robyn Longhurst, compiled by Max Oulton, 1999.

questions such as: 'what activities have you continued to carry out during pregnancy and what activities have you reduced or stopped carrying out during pregnancy?' and, 'which places have you continued to visit during pregnancy and which places have you reduced visits to or stopped visiting during pregnancy?' I conducted focus groups (see Longhurst 1996a), interviews and in-depth (ethnographic) work. I also spoke with some of these women's husbands/male partners, and with Hamilton midwives (for more information on the fieldwork and analysis of data see the Appendix).

One of the findings of this study of pregnant women was that many of the women increasingly withdrew from public space the more visibly pregnant they became. While some pregnant women continued as usual, most did not. Although model and actor Demi Moore may have appeared on the cover of the glossy magazine *Vanity Fair* naked – and eight months pregnant – in August 1991 (see Jackson 1993: 220–221) most of the women in this study would not dream of making such public statements (although see Longhurst forthcoming on the pregnant bikini contest in Wellington, New Zealand). There are many discourses in Hamilton which function to confine pregnant women to private space (see Longhurst 1996b).

## Early confinement

The root of the word 'confine' is from *con* – together + *finis* – end, limit, boundary. Confine as a noun refers to boundaries, bounds, frontiers or borders. The confines are the limits within which any subject, notion or action is restricted. Confine as a verb means to bound, limit, banish – to keep or restrain (a person) within their dwelling, to oblige them to stay indoors, or in their own room or bed. To be confined also means to be in childbed; to be brought to bed; to be delivered of a child. The period from the onset of labour until the birth of a child is commonly known as confinement (*Oxford English Dictionary* 1991: 312). Confinement came early for many of the 31 pregnant women in this study.

Confinement is a recurring image, not just in pregnant women's lives. Many women at various stages of their life cycle experience a sense of confinement (see Rose 1993b: 27). Angela, in describing the position of working-class 'white' women, claims:

> . . . if I have to think of one word that could work as a motif for this experience [of being a working-class white woman], it is confinement – the shrinking of horizons, the confinements of space, of physical and assertive movements within institutions, the servility that masqueraded as civility, the subjugation of my body, emotions and psyche.
>
> (Angela 1990: 72–73 cited in Rose 1993b: 27)

Marilyn Frye (1983: 4 cited in Rose 1993b: 27) reiterates this point claiming that women often experience 'being caged in: all avenues, in every direction are blocked or booby trapped'. Rose (1993b: 27) comments:

> Women cannot move freely, and this is not only a question of physical mobility; their writings resonate at the same time with frustration about who defines and delimits them. Angela chooses to chart her sense of restriction in a photo-essay of her body. This is significant because being defined by being looked at is central to this sense of confinement.

Morales (1983: 108 cited in Rose 1993b: 27), a Puerto Rican woman, makes the point that for black women part of their difficulty in negotiating some environments is having to look, act and sound white.

In my study of pregnant women in Hamilton there was plenty of evidence suggesting that pregnant women become increasingly confined to the home. For example, in response to my question about how things have changed since she become pregnant, 16-year-old Sarah, who was 25 weeks pregnant at the time of the interview, noted that she had stopped drinking alcohol, 'mostly stayed home', had a 'way boring' life and had a boyfriend who still went out with his mates to nightclubs and parties. A number of the participants also discussed withdrawing from sport, paid employment, restaurants, bars and cafés.

Pregnant women's withdrawal from public space can also be seen in symbolic maps[1] that they drew of their lifeworlds at the beginning of interviews/focus groups. I advised informants to draw these maps as though it was a 'parlour game' (Macdonald 1992: 14). As a result, conversation flowed freely as the pregnant women laughed and compared and contrasted drawings.

To return to Sarah, her maps (see Plate 3.2) were perhaps the most poignant example of a lifeworld that had became increasingly confined during pregnancy. From very early on in her pregnancy Sarah stopped attending parties and instead stayed home. She also stopped attending courses (her mother insisted that she give them up – 'Mum said that she doesn't want me to work any more. I just have to stay home and rest'). Netball, running and staying out late were absent from Sarah's sketch of her lifeworld during pregnancy. Sarah considered herself to be 'single' before becoming pregnant; after becoming pregnant, she became part of a 'couple'. On Map B Sarah writes 'Boyfriend being more aggressive'. While the change in Sarah's lifeworld represented in her two maps (a change that happened in just 16 weeks) may seem rather extreme, it was by no means unusual.

In Map A, representing Katie's lifeworld before pregnancy (see Plate 3.3), she includes, amongst other things: Friday afternoon gins, going out of town in her car each day (this was for her employment), boating at the beach and outings with women friends, such as going shopping and attending craft classes. On Map B,

*Plate 3.2* Sarah's symbolic maps of her lifeworlds before pregnancy and at 16 weeks pregnant.

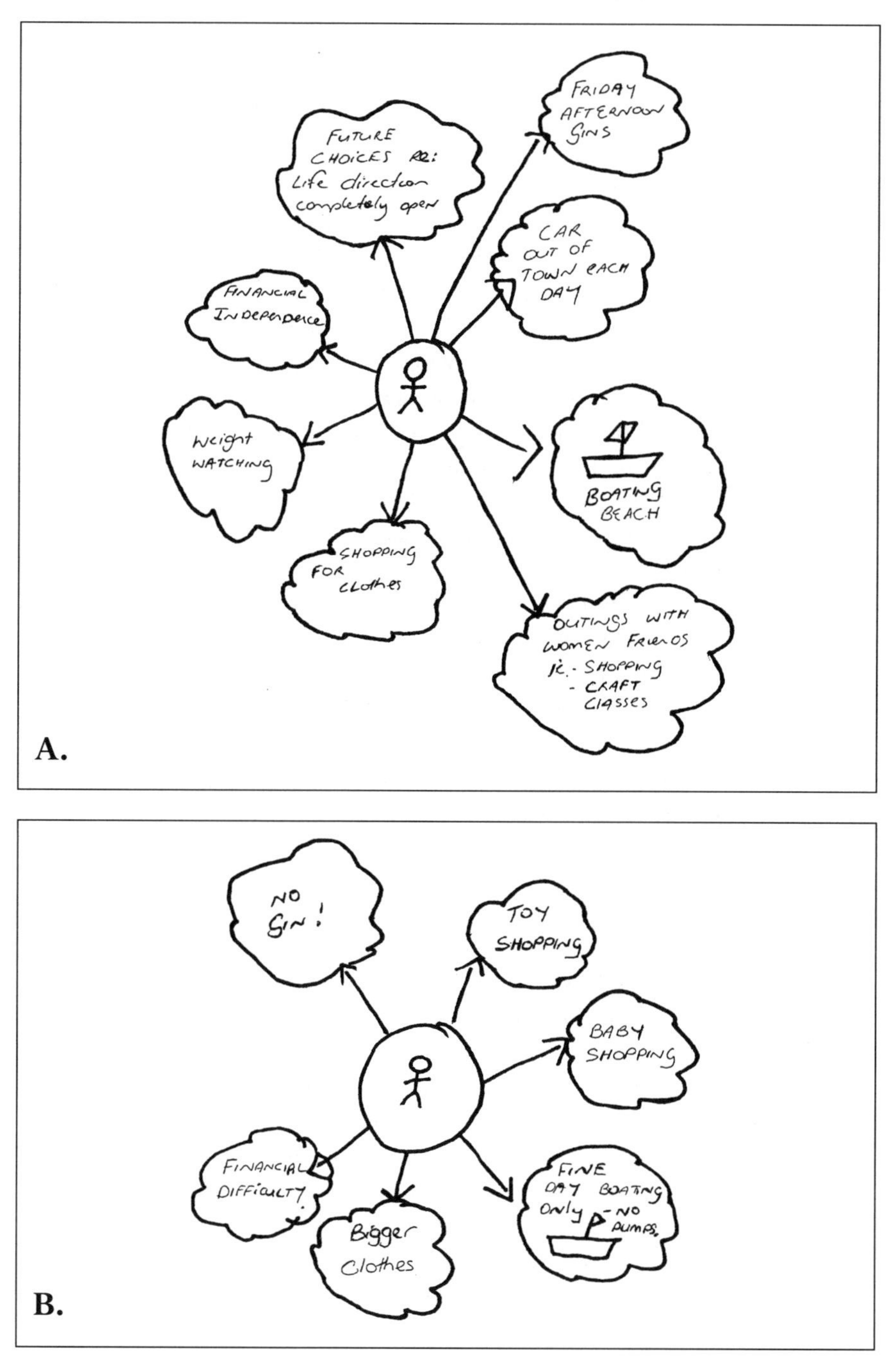


*Plate 3.3* Katie's symbolic maps of her lifeworlds before pregnancy and at 34 weeks pregnant.

which represents her lifeworld at 34 weeks pregnant, Katie includes no gin, financial difficulty, fine day boating only and toy/baby shopping. The maps indicate a marked change in Katie's lifeworld, that is, her activities and the environments she inhabits. While Katie and Sarah could be considered to be at opposite ends of a socio-economic scale (Katie's household income is over NZ$60,000 (approximately US$ 30,000), Sarah's household income is under NZ$10,000 (approximately US$5,000)), different in terms of their age (Katie is 34 years old, Sarah is 16 years old) and different in terms of their ethnicity (Katie is Pakeha, Sarah is Maori), their maps both indicate a withdrawing from the public realm during pregnancy.

I have mentioned just two of many maps. They indicate some of the ways in which most of the women seemed increasingly to orientate their daily activities around the home (and the coming baby) and withdraw from public activities. In including these two maps in the text I did not select only those that neatly 'fitted' my argument. In fact, on the contrary, I looked for maps that I could use to problematise, contradict and contest the notion that pregnant women in Hamilton tend increasingly to withdraw into 'domestic' space during pregnancy. However, none of the symbolic maps clearly illustrated this. For the majority of participants their public lifeworlds shrank during pregnancy.

One of the most commonly cited reasons for this withdrawal was physical discomfort in public environments. Some of the pregnant women claimed that seats were unavailable or uncomfortable, others noted the lack of toilet facilities in public spaces or the discomfort of smoky environments. However, simply changing these environments to make them comfortable would not necessarily lead to pregnant women having greater involvement and visibility in the public sphere. Built environments and power relations are inseparable, and change needs to happen at both the material and discursive levels. Changes to built environments alone would not necessarily mean that pregnant women would be able to visit swimming pools without feeling self-conscious, or that they could walk along a beach nine months pregnant in a bikini and feel confident about their 'image', or that they could go surfing, snow skiing or mountaineering without being encouraged to feel guilty about risking the well-being of the fetus. Changes to built environments would not be sufficient to ensure that women could be employed through their pregnancy without being encouraged to feel that they do not adequately reflect the corporate image or that their brains have 'turned to jelly' (*Waikato Weekender* 1991:14).

It is too simplistic to claim that the decision of the participants to withdraw from public space is due to the physical, material, corporeal demands of pregnancy. It is undeniable that the material body of the pregnant woman is different from the material body of the non-pregnant woman. The pregnant woman is likely to be 9 to 14 kilograms heavier, she may be retaining some fluids, feel tired, have swollen feet, varicose veins, pain in her groin, shortness of breath, backache and haemorrhoids. Some women, on the other hand, report feeling energetic and healthy for the duration of their pregnancy. It is not the weighty, material body in discomfort or health, not simply biological bedrock that can explain pregnant women's withdrawal from public space. The material pregnant body is simultaneously constructed and inscribed by the discourses that surround pregnancy. It is these discourses that are too frequently ignored in understandings of pregnancy. Therefore, in the reminder of this chapter, in an attempt to explain some of the complexities surrounding the participants' withdrawal from public space, I discuss one of the discourses that inhabits pregnant bodies, that is, pregnant bodies as 'modes of seepage'.

## 'Modes of seepage'

In this section I draw on Elizabeth Grosz's (1994a: 203) work on dirt, sexual difference, body fluids and the inscription of women as 'modes of seepage' outlined in the previous chapter. I also develop Young's (1990a: 145) notion that specific bodies become 'culturally defined' as 'ugly'. Pregnant women's body fluids pose a threat to social control and order. Their border ambiguity may become, for others, a threat to their own borders and they may react with feelings of loathing as the means of restoring the border separating self and other. They may try to confine the pregnant woman in the private realms because of the threat that her leaking, seeping body, her womanly 'corporeal flows' (Grosz 1994a: 202) and her splitting self, poses to a rational public world. Dorothy Dinnerstein (1976 cited in Silverman 1989: 7) proposes that 'the fear of mortality is associated with contact of one's own body with the flesh of the mother's body in all its delightful and terrifying ways'. Silverman explains that the memories, vulnerabilities and desires of infantile experience are segregated and associated with mother, home and personal family life. Men tend to cut off from these and do not want to be reminded of them in the public world of politics, economics and foreign affairs.

Grosz's (1994a: 203) thesis that 'women's corporeality is inscribed as a mode of seepage' is a particularly useful and rich idea to consider in relation to pregnant bodies. Although Grosz (ibid.) discusses women generally she does not make any specific reference to the pregnant body as a 'mode of seepage'. There is potential here to build on her argument. In terms of the mind/body dualism, if men have been defined on the side of the mind and women on the side of the body then there are certain bodily zones that serve to emphasise both women's difference from and Otherness to men. Pregnancy, both culturally and biologically, poignantly marks this sexual difference. The enlarging of the breasts ready for feeding the infant, the swelling of the stomach, the threat of the body leaking fluids and splitting itself into two, all this marks women's sexual Otherness.

Bearing the status of a leaking, secreting embodied Other plays a crucial role in pregnant women's withdrawal from public space, which is, after all, the realm long associated with Rational Man whose body is considered to be solid and in control. Pregnant women's bodies are inscribed as 'modes of seepage' in numerous ways with their waters breaking, morning sickness[2] and increased urination. I discuss each of these 'modes of seepage' or 'corporeal flows' in turn.

Sheila Kitzinger (1989: 228), one of the foremost pregnancy and childbirth educators of the past two decades, explains that:

> When the membranes surrounding the baby have been pressed down like a wedge in front of its presenting part (usually the head) and pressure has built up, the bag pops. It may do this suddenly with a rush of water or, and this is more likely, with a slow trickle.

Kitzinger (1989) advises 'you may not be quite sure whether the bag of waters has burst or if you are wetting your pants'. This was a common topic of conversation for the Hamilton participants.[3]

ROBYN: A number of women have said, sort of almost jokingly to me, they have fears about their waters breaking – like when you pop out to the supermarket or something. Suddenly your waters break and you go into labour. Have you thought at all about that?

DOROTHY: When I came here today actually [laughter]. It's the first time I've thought about it [laughter]. I actually put a towel in the car. It really is the first time 'cause um, my midwife reckons that for her, um, having your labour start with your waters breaking is fairly rare, you know, contractions start first and then you've got some sort of warning and can be prepared . . . I thought I'm overdue, it may be only one day but I am overdue. I'd better take a towel [laughter]. But ya just don't know. Like you say you can lose two drops and not even realise your waters have broken or you'll loose two cups and you'll know about it, so, I'm here, at your risk, sitting on your chair [laughter].

Conversations such as this often emerged when I was in a public arena (for example, shopping in Hamilton's CBD or at a supermarket) with a woman or women who were nearing the end of their pregnancy. The conversations were often lighthearted and the prospect of the woman's waters breaking while she was out in public was commonly laughed or joked about. I sensed, however, that this fear was 'real' and that the jokes and laughter about the prospect of it happening were a way of dealing with feelings of nervousness about an event that could prove potentially embarrassing.

Waters breaking signify a body that is 'out of control', and since bodies are not supposed to be out of control in public environments this is dangerous. While menstrual flows can usually be 'dealt with' by way of tampons or sanitary napkins, and there may be some warning as to the onset of menstruation, the flow of the waters breaking may be very sudden and involve a large rush of fluid that cannot be absorbed by a sanitary napkin. There seems to be an idea amongst a number of pregnant women that the waters breaking is a degrading physical process – it is a dirty process that involves getting rid of waste products from the body – and that this ought not to happen in public. It may also be seen as a kind of sexual act in front of other people.

Also at the start of labour a 'show' may appear. The word 'show' refers to (and is a euphemism for) a 'blood stained mucus discharge' that becomes apparent when the cervix begins to stretch (Kitzinger 1989: 228). Until the start of labour this mucus acts 'as a gelatinous plug in the cervix, sealing off the uterus' (ibid). Kitzinger (ibid.) explains that the 'gelatinous plug' can come out any time between about three weeks prior to the woman going into labour to when she is well advanced into labour. It is unlikely that this 'show' will be noticed by anyone other than the pregnant woman herself but it is another example of the leaking and seemingly out of control pregnant corpus. It is also an example of the self-loathing and discomfort that pregnant women can sometimes experience in relation to their own bodies.

When I began the project on pregnant women in Hamilton my focus was on women who were visibly pregnant. However, as the research moved into the final year and I myself became pregnant I was forced to rethink many of the issues that I was writing about. This included a move to consider the experiences of women in the early stages of pregnancy, and in particular, their experiences of morning sickness. Whelan (1982), who has conducted research within a North American context, shows that 50–75 per cent of all pregnant women experience morning sickness. Morning sickness can occur at any time of the day or night and recur for months. For many women it is not the small nuisance of early pregnancy that society in general perceives it to be. During my interviews and focus groups in Hamilton many of the respondents claimed that they had experienced some form of morning sickness.

For some, morning sickness signalled to the woman herself, and sometimes to others, that she was in fact pregnant.

ROBYN: So you found out you were pregnant?

SONYA: Yeah, on tour [with a band of musicians] . . . It was actually kind of a harrowing experience really 'cause I couldn't tell anyone. I hadn't told my boyfriend and I had morning sickness and they all thought it was travel sickness and they were giving me travel sickness pills. And um, one of them guessed [laughter] when I didn't take the travel sickness pills.

Many of the women discussed morning sickness in terms of their experience of paid employment. Dorothy, who was working as a sales representative when she became pregnant, explains:

DOROTHY: I got morning sickness [laughter] which was actually afternoon sickness, so yeah, it was a big change.

ROBYN: So you, while you were having that afternoon sickness, you were still working?

DOROTHY: I was still working.

ROBYN: How was that?

DOROTHY: Hard. It was really hard. Um, in a way I preferred afternoon sickness to morning sickness because it meant I could start the day with a good breakfast. I always managed to get in a good breakfast, but um, by lunch time we had this thing where we all sat round and had lunch together and I just couldn't stand the sight of anyone eating. I couldn't eat myself. The only things actually, they called me the mandarin queen, 'cause that was about all I could eat [laughter] was mandarins. And then, come meal times I just, my husband cooked for about two months solid. I was, because I just couldn't face the smells. I just picked [Robyn: Right] and I found that by two o'clock in the afternoon, I was just ready to go to bed. I was history, really tired, so I found I tried, at work I had to try and get all my important things done in the morning and I usually found that I was in the office [rather than visiting clients] the entire afternoon, which they didn't really like but understood.

Christine, who was working as a bank teller when she became pregnant, says:

> The first few months were really quite hard yacker. I was sick for about five weeks. Yeah, I would have been four months when I left work . . . They probably thought that this teller doesn't look very fired up. She looks rather pale . . . They [the other bank tellers] were pretty good. The bank was good except I wasn't very impressed when my accountant, who's in charge of the staff, she said that if I kept being sick that I might have to take my holidays earlier. And, I'd been in the bank five years and I'd hardly used any of my sick leave and I was due for about two weeks holiday and she said if you keep being sick it might be a good idea to move your holidays forward and I wasn't very impressed that she even suggested that because I mean if it was the flu or whatever you'd take two or three weeks off so I don't see any difference I thought that was really slack.

Jill, an office worker, claims that she herself had not been affected dramatically by morning sickness but told a story about a friend.

> One of my friends, she's a teacher . . . she'd get up, be sick in the morning, go to school and then come home and be sick. She wasn't sick so much during the day. So I think that can be quite sort of stressful you know, having to be sick and still going to work.

The body that threatens to vomit is not a body that can be easily trusted to occupy the respectable public realm, including the workplace. The pregnant woman who enters the public space risks 'soiling' herself and perhaps even others with matter produced by her body. Her body threatens to contaminate and to pollute; her bodily 'difference' becomes evident in workplaces, and various other public places. Based on 248 returned completed questionnaires from New Zealand women who had morning sickness, Birks (1993) reports 82 per cent claimed that 'unsympathetic attitudes' were a problem. Penny, who worked in a government department, says:

> There's another lady I worked with. She said when she was pregnant she was sick quite a bit and once or twice got off the bus and was sick into the gutter or something like that, you know, it would be revolting if you were like that.

Being sick into the gutter is not socially acceptable behaviour, and it would appear that sympathy is lacking. It signals a body that is 'out of control' and in need of confinement. It is possible to identify other bodies that are also sick into gutters, particularly small children who get car sick (they are too young to discipline their bodies) and people who are drunk (they are too contaminated by inebriating liquors to discipline their bodies). Both these bodies – the child's body and the 'drunk's' body – are subject to effective confinement by sanctioned agents of control such as parents, caregivers or the police.

My own experiences of nausea (which occurred in the morning, late afternoon and evening) in the first three months of pregnancy also testify to a withdrawing from the public space. I withdrew from my workplace – the university – as well as recreational space such as restaurants, bars and so on. At the university I have my own office (although many women do not – see Spain 1992: 199–230) and was therefore able to hide a plastic bucket under my desk 'just in case' I needed to vomit. Fortunately the problem of where and how to empty the bucket did not arise but I was troubled that it might. How would I hide this vomit from colleagues? Did I need to? How might they respond to matter that had formerly resided inside my body making its way to the outside? Would they respond with abject horror and attempt to send me home? I did stay at home on some occasions in order to avoid these problems despite the fact that I did not have the resources at home to carry out my work effectively.

Morning sickness affects not only women's activities in relation to paid employment but also in relation to many other activities including grocery shopping.

MARGARET: I had a lot of trouble. I had really bad morning sickness right up until basically from week six until the end of the first trimester and um like it was an effort to do anything. Like when it was really bad I could be throwing up every ten minutes and um about all I could eat were raw carrots, figs, prunes, and um fruit . . . I've always been, you know, a bargain hunter... but I found when I was feeling really sick I just [went] into the supermarket and I'd just skirt round as fast as I could and . . . I would just grab... and it would just be a matter of trying to not throw up while I was waiting in the queue and get through and get out to the car.

Morning sickness also affected Margaret's life as a university student, especially in relation to participating in geography field trips. I asked Margaret how her life as a student had changed since she had become pregnant.

MARGARET: When I was really really sick with morning sickness it was just right at the very worst part, was right at the very end of the first term and in the last six days of the last three weekends. I had um field trips on five of those days. One of which was an overnight and I had tests and essays and everything all due at once and I was just feeling just so grotty. And that was really hard, and um, like I had one particular um field trip that was compulsory for 'Coastal Processes and Management', which I really didn't think I'd be able to do. But anyway, in the end I got into it all and, I sort of talked to him about it and to Professor H . . . and asked him if I could take my own car and just follow the bus and then if I felt sick at least I could just stop. But he wasn't too keen on that.

ROBYN: Did he say why?

MARGARET: Because he normally talks in the bus and I would have missed that information. But in the end um I didn't really want to do that either 'cause I thought I'd be too tired with the driving as well 'cause it's pretty much all day. You know, we left here at eight o'clock, we didn't pull into the camping ground until sort of 5.30 at night, or something like that. And in the end my husband said to me 'look it just doesn't matter where you are just get off the bus, just get off the bus and give me a call and I'll come and pick you up'. The field trip was in the Bay of Plenty. So that actually made a really big difference, 'cause then I didn't feel that I was sort of trapped on the bus for the whole weekend and if I did just start to throw up the whole time I could at least you know, there was an escape. But I was really uptight about that for a number of weeks beforehand. You know I was thinking I couldn't cope. Most of the other field

> trips I went on I just arranged to meet them at the place and then I would just drive 'cause I found that even um, you know like normally you don't get carsick. Like I get carsick as well, but this was quite different. Like normally I never get carsick when you're driving, but I found I was actually dry retching when I was actually driving home from varsity and things like that – you know it was really bad.

Clearly morning sickness affects women in relation to many activities in their lives.

In addition to the waters breaking, a 'show' appearing and morning sickness, pregnant women's enlarged breasts can also be understood in terms of seepage, viscosity and fluidity. In the course of conversations with pregnant women many mentioned changes in relation to their breasts. Perhaps this is not surprising given that nearly all popular advice manuals on pregnancy contain sections on breasts. These sections usually contain sub-sections on brassieres, changes in early pregnancy, engorgement, inverted nipples, sore and cracked nipples and tender breasts.

In general the conversations that I had with women about their breasts seemed to focus on the enlarging of their breasts right from the first days of pregnancy. One respondent, a dance teacher who was aged 30–34, claimed that she enjoyed having larger breasts.

MICHELLE: One of the really nice things is I've got breasts. I was always one of these flat-chested people and so I feel so voluptuous during pregnancy.

Most, however, did not like the changed form of their breasts.

DENISE: They look alright when they're in the bra but when they're not in the bra [screws her face up].

KERRY: Saggy-baggy . . . I get the veins and Denise gets the *big tits* [laughter] [inaudible].

ROBYN: I got my sister to sew me this bikini top with a kind of bra inside it 'cause it, that was helpful (//)

DENISE: (//) Yeah right . . . whenever I've had to buy togs, I've gotta um buy ones with underwires anyway 'cause they either flatten ya off, or they don't support you . . . They're just that stretchy material and ya put them on and . . . they either just make ya boobs look all floppy or they really flatten them.

A little later in this same conversation Denise describes her breasts when she lies down.

DENISE: They are really uncomfortable, really, like I lie on my side (//)

KERRY: (//) I don't feel so bad now.

DENISE: And it feels like it's on your back and you're lying on it and I have to keep going like this [cups the outside of one of her breasts in her hand and rolls it to the front] and then try and get like that if I let the boob go before I lie back down on the bed, it rolls around the side [laughter]. It's really horrible – they come out to the sides.

Denise's comments attest to the increased fluidity of her now enlarged breasts.[4] It is evident from her comments that she does not feel altogether positive about this change in her corporeality – 'it's really horrible', she says. Denise does not like the way that her breasts now 'come out to the sides' when she lies on the bed on her back.

Grosz (1994a: 205) notes that: 'The fluidity and indeterminacy of female body parts, most notably the breasts . . . are confined, constrained, solidified through more or less temporary or permanent means of solidification by clothing or, at the limit, by surgery'. While pregnant women do not 'firm up' their enlarged breasts through surgical implants, they are however advised that 'from the first days of pregnancy, you will need a bra which gives good support' (Kitzinger 1989: 130). Kitzinger also advises that: 'Heavy breasts, allowed to hang without support, may develop stretch marks . . . which will leave you with silvery streaks after the pregnancy. A woman with large breasts may prefer to wear a lightweight bra at night too during pregnancy'. Even in the privacy of one's own bed there is a constructed need for the breasts to be confined within a bra in an attempt to control and solidify them.

A number of the Hamilton participants discussed their feelings towards their enlarged breasts in relation to wearing swimming costumes. This is not surprising given that when wearing a swimming costume the size and shape of breasts is very noticeable – pregnant women are likely to feel subject to public gaze. Pregnant women may also not have adjusted to the way their 'new' breasts feel. This was the case for Ngahuia.

> That [large breasts] was an issue really because my normal pair of togs I found they dragged your bust down whereas, they were comfortable around the *puku* [stomach] but dragged your breasts down and I didn't like the feeling of it, and I didn't like how it looked so I thought I'm going to invest in a good pair of maternity togs which hold you up here.

Note that Ngahuia wanted her swimming costume to hold her up – to support and solidify her breasts.

Similarly Denise says that when she ran across the road in downtown Hamilton she was concerned that onlookers would see her 'boobs' moving. Denise is of slim build and was 27 weeks pregnant at the time of this interview.

> I was running across the street today in the rain 'cause it was raining and we had to run across the pedestrian crossing and I was running along and your boobs bounce up and down . . . I was concerned 'cause I thought there were people parked in cars waiting for the lights and I was running along. I knew they'd bounce up and down.

Another point to note in relation to the way in which women's breasts change during pregnancy is that the body produces colostrum. Colostrum is the earliest form of milk. Kitzinger (1989: 350) claims that it is 'rich in protein and an ideal first concentrated food' for the new born baby as it lines the baby's gut with a protective layer against bacteria. Colostrum may also provide the baby with antibodies to diseases that the mother herself is resistant to. From when a woman is approximately 28 weeks pregnant 'colostrum may leak from . . . [her] breasts' (Kitzinger 1989: 375). Paula noted in her journal that her breasts during pregnancy were 'larger and leaking'.

So far I have discussed how the volatile pregnant body in Hamilton curtails many public activities and becomes increasingly confined to private spaces. Public spaces do not accommodate bodies that are understood to be and inscribed as 'modes of seepage' – bodies that vomit, have a 'show', breaking waters, a 'rush', a 'trickle', enlarged and leaking breasts. I have not discussed the tendencies in pregnant women to cry more, to need to urinate more frequently, to sweat more with the effort of carrying the extra weight of the pregnant body or the threat of the baby actually 'spilling out' – being born – in a public place (other than in the hospital). Grosz's (1994a) thesis that women are constructed as 'modes of seepage' could also be used to examine these aspects of pregnancy. But not only are pregnant women's bodies inscribed as 'modes of seepage' – bodies whose leaky boundaries are to be feared – but the discourse is further compounded by descriptions of their bodies as 'ugly' and abject.

## 'Ugly' and abject bodies

Young (1990a: 145) argues that some groups come to be defined as ugly and abject. Consider the following example. Kerry recounts a conversation that she had with boy aged 14. The boy's feelings of abjection, and his representation of the pregnant body as ugly, are clearly revealed in the conversation.

> I was reading out the um week by week thing [a summary calendar of the various developmental stages of the fetus] to this fourteen year old that comes and milks with us – Mark. And um [laughter] I read out week twenty... and it said somethin' about um 'in week twenty your navel should pop out at any stage from now on . . .' He's going '*O o o yuck*' he said

> 'Oh you're going through a real ugly stage now Ker', and I said 'bloody hell' [laughter] and then I was in the shower this morning and I've got all these veins comin' up on my legs and I'm thinking 'Jeez, I'm going to the pack, I'm going [laughter] 'Oh Jeez' and 'Oh I'm horrible'. And he said 'Oh my aunty . . . I saw her just about two days before she went in' and he said, 'it's so ugly' [laughter].

Kerry's male colleagues described the ultrasound scan to her in the following way: 'For 15 to 20 minutes you lie there with all this gel all over you with your *guts sticking out*' (my emphasis). Clearly his use of the phrase 'guts sticking out' to describe Kerry's pregnant stomach is interesting in that it does nothing to indicate the attractiveness of her stomach, rather, to the contrary. Another male colleague informed Kerry after she had spilled a cup of tea on herself 'when you start breastfeeding you'll have more stains than that over your body probably . . . a good tip is to use your husband's hankies and put them down you'. Yet again, the comments indicate a body that leaks and that will be covered in unwanted bodily fluids – stains – which will need to be controlled.

Paula recounts a story of a friend visiting her at home. She explains that at about 22 weeks pregnant, a friend, who she sees about once a fortnight, came to dinner. As Paula opened the door to her, her first exclamation was 'Gosh, you get worse every time I see you'. Paula explains that in this instance, 'get worse' was equated with looking larger. Paula's pregnant stomach had become more evident since she last saw her friend.

Members of groups subject to oppression 'often exhibit symptoms of fear, aversion, or devaluation towards themselves and toward members of their own groups' (Young 1990a: 147). In relation to pregnant embodiment, it is useful to examine the words and phrases used, not only by others but also by pregnant women themselves to discuss their own embodiment and the bodies of other pregnant women. Images of fatness, disability, incapacity, discomfort, and ugliness are abundant.

'We were gonna park in the disabled car park', says Denise. Denise's statement makes a connection between being pregnant and being disabled. Christine claims: 'People sometimes treat you like you're just about handicapped when you're pregnant'. Ngahuia, although much less directly, also identifies a link between pregnancy and disability. 'I use the paraplegic toilets', she says.

Pregnant women often discussed their perceptions of their bodies when I asked whether they had any photographs of themselves pregnant. When I asked Mary Anne, who was 36 weeks pregnant at the time of the interview, this question, she replied:

> *No* and I don't intend to have any... I don't want anyone to look at my big bum. I don't mind my body shape of the baby, it's my hips and thighs that I don't like the thought of looking at.

Jude, a university student who was 33 weeks pregnant, responded similarly to the same question.

> I never wrote a diary as a child in case somebody that I didn't want to found it and I think it would be the same with a photo. Somebody would be leafing through my album and laugh. Somebody that I didn't want to see me like that.

The judgements passed on pregnant women's bodies by others may lead to them feeling uncomfortable about their own bodies. For example, Terry's mother-in-law left Terry feeling as though she 'couldn't win'.

TERRY: She [her mother-in-law] would cook this nice dinner and you'd have to eat it all up to please her . . . there's always this constant thing with her every time she sees you 'Oh, are you eating enough?' and I'd say 'Oh yes'. I'd say 'Oh look, you should see how much I've been eating, look at this weight I've put on', and then she'd say 'Oh, oh dear I think you should be exercising'. You know and you think 'you can't win'. And then she'd say 'Look . . . [Terry] I think you're too big, you know. Look at – 'cause her daughter is ten weeks further on pregnant than I am and she's put on the same amount of weight as me – she's really put on very little for how far along she is. And I've found that really difficult, and then I'd go around say to the daughter's place and she'd say 'Oh gosh, look at you, you've put a lot of weight on your face haven't you?' [laughter] And you end up, you know, you're self-conscious enough about it.

MARGARET: Yeah, you are, it makes it worse I think.

TERRY: I would actually probably not have thought about it unless she'd said something about it. And then I went to my doctor and I said 'have I got on too much weight?' I said, 'you know, my mother-in-law says I'm too big you know'. And I got really paranoid about it.

MARGARET: I did at first too about it.

TERRY: Yeah, I found that quite difficult, especially like in the last two weeks I've put on a kilo each week, you see and I think 'oh I've put on ten already and I'm 25 weeks so, what am I going to be like at the end?'

One of the topics of conversation that tended to lead women to talking about their pregnant bodies was swimming and swimwear. General practitioners often advise pregnant women that swimming can be a useful activity to pursue in order to keep healthy. Yet it is seldom made explicit that swimming usually involves pregnant women revealing their new body shape in public. Given that many pregnant women dress in baggy garments that act to disguise their swelling stomach – 'you do cover up your pregnancy. You always wear big clothes that are bigger than your tummy', says Jude – it comes as no surprise that many felt uncomfortable wearing a swimming costume and even decided not to swim.

ADRIENNE: I was a bit embarrassed the first time I went swimming. I sort of thought – oh, you know! But, it was comforting to see other pregnant women there . . . 'cause people tend to accept that when they see more pregnant women there – it's all right . . . the very first day I went there wasn't a pregnant woman in sight. I went in the afternoon and there wasn't a pregnant woman in sight and you feel as if the whole world is watching you.

JOAN: When I was about five months . . . we went swimming, I was swimming in the pool, but lots of the girls, because I wasn't that big, had bigger tummies than I did anyway, so it didn't bother me at all but I don't think I would swim now. I'm too uncomfortable – people looking at me.

MOIRA: I went swimming yesterday at Te Rapa pools. It doesn't really worry me only that Dave [the interviewee's husband] works there and so I'm maybe a bit more self conscious because they know me rather than if I just didn't know anybody there I probably wouldn't worry as much.

None of the visibly pregnant women I interviewed wore or had considered wearing bikinis while swimming in public (although a number of them had worn bikinis prior to pregnancy). Yet it was evident that women who did swim when pregnant faced problems in obtaining swimwear which fitted them comfortably.

DAWN: I actually bought some maternity togs because they were on special [sale] . . . they are really expensive. I tried on a normal pair and they just didn't feel right.

ANGELA: I just loaned a pair [of maternity togs] off a friend yesterday that are black so that is quite good, covers me up a bit. I probably wouldn't wear black togs if I wasn't pregnant.

There is an uneasiness about the public exposure of pregnant bodies. This uneasiness is often on the part of both the pregnant woman and on the part of those her view her. Howard (Christine's husband) talks about his response to seeing pregnant women on beaches.

> I've been to a lot of beaches . . . and I've seen pregnant women in all sorts of states of dress and undress and it seems quite a normal, ordinary course of events. Yet it is different . . . I guess something in me tells me that a pregnant woman is somehow in a different status to a non-pregnant woman, in a way a pregnant woman is sort of non-sexual, outside of

> courtship rules. In the meat market – beaches do feel like that sometimes – there are lots of participants who are not involved in the game, for instance families and pregnant women.

Howard raises an important point. Pregnant women are often perceived as being 'outside of courtship rules', they are constructed as 'non-sexual' beings despite the fact that at the same time they are clearly marked as having been sexually active. Yet once a woman is pregnant she is often considered to be no longer sexually available, active or desirable, even though her own desires may have increased. Young (1990b: 166) argues that the pregnant woman's

> male partner, if she has one, may decline to share in her sexuality, and her physician may advise her to restrict sexual activity. To the degree that a woman derives a sense of self-worth from looking 'sexy' in the manner promoted by dominant cultural images [in Hamilton the dominant culture defines feminine beauty as slim] she may experience her pregnant body as being ugly, and alien.

Constructions of the pregnant body as ugly, alien and not 'sexy' or sexual help to explain why the pregnant body is so often considered to be private and in need of concealment. It also helps explain why such an enormous furore emerged in Canada and the United States when actor and model Demi Moore broke these unwritten rules and appeared on the cover of the glossy magazine *Vanity Fair* naked and eight months pregnant. While readers of the magazine were accustomed to seeing images of naked and semi-clad women they were not accustomed to seeing such women pregnant. Images of motherhood (read: not sexually available) and supermodel (read: sexually available) became uncomfortably entangled. The cover, and the photographs inside, caused such strong feelings amongst so many Canadians and North Americans that it was treated almost as pornographic material. The magazine was withdrawn from some supermarket shelves and sold in a protective plastic wrap in others (Jackson 1993: 221).

Whether men experience pregnant bodies as ugly, to be feared and abject, to a lesser, similar or greater degree than women, is difficult to determine. There is little doubt that women are in no way exempt from feeling abjection towards the pregnant body, but from comments made by some of the Hamilton research participants it is possible that men may find the uncontainable, seeping corpus more difficult to 'deal with'. Rebecca, who was aged over 35, explained that not only was her husband going to accompany her during the labour and birth but so too was a woman friend. Rebecca explained 'I've got a husband that's a bit panicky' and 'he'll probably faint'. The friend's presence was for her husband's sake – 'it's for

Stewart' – rather than for her own. Another research participant, Katie, whom I spoke with again after her birth, reported that her husband had fainted during the labour. He fainted not at the moment of the birth itself but when Katie was receiving an epidural injection in her back (not when the baby penetrated the bodily boundary from the inside but when a needle penetrated the bodily boundary from the outside).

Kerry explains that at antenatal class when watching a video of a woman whose waters had broken and who was going into labour one of the husbands had to leave the room – 'he was being sick', said Kerry.

KERRY: Yeah, he walked out.
DENISE: One of the fathers watching it?
KERRY: Yeah [laughter] he was gone [laughter] . . .
DENISE: He's gonna be a lot of use isn't he? Carl won't like those videos.
KERRY: It does, it puts them off.

Later on in the same class, the pregnant women and their husbands were informed that during labour the pregnant woman could pass a bowel motion. On their way home from the class Kerry's husband said to her 'I hope you don't shit yourself'. Kerry responded by saying that if she did he couldn't blame her. 'It'd be horrible', her husband responded. Kerry's husband was disgusted by the prospect of what Kerry's body might do.

Yet it was not only men who responded to, and constructed, the pregnant body as ugly, to be feared, and abject. There may well be some differences between men's and women's responses to the pregnant body – men may be more afraid than women of the body that threatens to seep and split its one self into two – but most of the women themselves also constructed their pregnant bodies as ugly.

While some respondents did report feeling good about the baby kicking and about a sense of fascination concerning their changing body shape, most of the comments about their corporeality were negative. In fact, of the 31 pregnant women with whom I spoke, only two reported feeling really positive about their body shape. Michelle, a dance teacher, claimed: 'A pregnant body is really quite beautiful, it is just the feeling of, I don't know, it's like I feel good about being pregnant.'

In response to the question 'how do you feel about your pregnant body?' Ngahuia, a university lecturer, replied:

> I love it. I think it's good to have a positive attitude about it because so many women get put off by the fact that being pregnant means your body changes and you look awful, but let's face it, it is natural to look like that when you are pregnant and it's good to have a positive attitude and to set positive examples to other women to encourage them.

While these accounts of pregnant embodiment might apply in similar ways to all women who are pregnant, at the same time the group of 31 women with whom I spoke represent only 'one slice' (Young 1990a: 141) of the oppressions of racism, sexism, homophobia, ageism and ableism. The accounts I have offered attempt to explain how pregnant women in Hamilton in 1992–1994 have become culturally defined as abject, seeping and ugly bodies, but conceptions of bodies do not stay static over time and place. Abjection is a result of complex, culturally and temporally defined social constructions of the pregnant subject. There is nothing intrinsic to the biological formations of pregnancy that requires a loathing of pregnant embodiment.

## Containers and containment

Given that pregnant bodies are constructed as materialities that are abject, dangerous and to be feared, then it is not surprising that they are also considered to be bodies that need to be contained and controlled. There are many ways in which attempts are made to contain pregnant bodies. First, the fetus is often treated as though it were a public concern. Pregnant women's rights to bodily autonomy are considered to be questionable. Pregnant women are often treated as though they are little more than containers for unborn children. Second, this leads to pregnant women's stomachs being subject to public gaze and often touch. Their 'bodily space' is frequently invaded. Third, pregnant women are often constructed by lovers, husbands, partners, friends, family, strangers, health workers and themselves as being in a 'condition' in which they must take special care in order to protect the well-being of the fetus. Fourth, pregnant women are subject to dietary regimes in an attempt to control what enters their bodies.

While the Cartesian separation between mind and body underlies all western understandings of embodiment, it is possibly even more evident when considering pregnant embodiment. Pregnant women often comment that attention and conversation seems to rest continually on their embodiment even to the point where eye-contact is not made with the pregnant woman and instead glances are directed at her stomach.

MARGARET: I went to town a couple of weeks ago and I noticed that I felt like every fifth person, instead of looking at my face, they would be looking at my stomach as I walked along.

When I accompanied two research participants, Denise and Kerry, to downtown Hamilton to go shopping for clothes, they both commented as we walked along the street that they felt people were looking at their stomachs. While one research participant found this experience frustrating, the other was not bothered by it and explained that she thought that looking at a pregnant woman's stomach is an 'unconscious thing' and that she probably does it herself.

I was asked in the course of this research whether I thought that people actually (really) look at pregnant women's stomachs or whether pregnant women just think or imagine that they do. My experience of being pregnant, accompanying pregnant women in public places, and my own glancing (gazing) at pregnant women's stomachs has led me to believe that people do actually look at them. It is not just a pregnant woman's imagining. This is not surprising since the abject is not only 'dread of the unnameable' it is also 'fascinating, bringing out an obsessed attraction' (Young 1990a: 145). Yet even if this gaze upon their bodies is imagined it could be argued that the effect is similar – pregnant women feel under scrutiny and act accordingly.

Many pregnant women also report people touching their stomach – not just loved ones or very close friends but also people who would not usually consider touching their bodies. On several occasions when I was working as a university lecturer and my pregnancy was clearly evident, students (whom I did not know very well) touched my stomach. It is not uncommon for some people to take the liberty of placing their hands upon pregnant women's stomachs. While one or two women claimed that they found this frustrating and disempowering, most appeared not to mind, and some even enjoyed it.

ROBYN: Have you found people have touched your stomach since you've been pregnant?

TERRY: Yeah

MARGARET: Some people are probably a bit – they don't know whether to or not but . . . [inaudible]

ROBYN: And do you mind it?

TERRY: No

MARGARET: I'm really used to it. Kevin does it all the time but I mean that's different, any but um. No it doesn't worry me.

ROBYN: Did it worry you, Terry?

TERRY: No, no I quite liked it actually. Just the other day when we went round to [her husband's] sister and brother-in-law, [name], they've got a daughter who is 10 and she straight away reached across, and I thought that was quite nice . . . Walter's Mum and Dad, they would like to, but they haven't. When I'm round there and it's moving I'll ask them if they'd like to. But at the moment, it's never moved while I've been there. But I think they're probably dying to have a little touch [Robyn: Yeah]. Yeah, but children especially.

The responses from two other participants in another focus group were similar.

ROBYN: Have you had that kind of thing happen? [people not meeting your gaze at eye level but rather looking at your stomach].

DAWN: Yeah [laughter]

ANGELA: It's 'oh you're pregnant, oh your boobies have grown' and I'm 'oh I know and, leave me alone' you know? Mm course I am, I'm having a baby. You feel like saying 'don't state the obvious' . . . But they do. They don't look at your face – it's straight, [touches her stomach] and it's like you feel, yeah, I feel like that I don't mind people touching my stomach. There's some people I don't like, but I don't mind.

ROBYN: People have done that to you?

ANGELA: Oh yeah, they've talked to me and, and you can see the hand waving. Oh go on then you know [laughter].

ROBYN: Have you had anyone touch your stomach?

DAWN: Yeah, but you know that's been, that's been okay for me. I don't mind.

Similarly, Christine had experienced people touching her stomach but did not seem to mind.

ROBYN: Have you found people comment on or touch your stomach?

CHRISTINE: Yeah, I think people do think they have sort of a right to sort of comment, but most people have been pretty good. It was hardcase like a friend the other day at church, he sort of gave me a little pat, I didn't mind, but it would just depend who it was type of thing.

Sonya, who was 26 weeks pregnant at the time of the interview, was less receptive to people touching her stomach although she herself admits to having touched other pregnant women's stomachs in the past without invitation.

ROBYN: Have you had people touch your stomach?

SONYA: Mm, not many though, which is good 'cause um I guess, it's not. I used to do that to other women, I used to go up and put my hand on as if it was my property because it stuck out [laughter] but it's not you know, and I know now how personal it actually is. It's nice to have somebody come up and put their hand on as long as it's invited on. I wouldn't like somebody just to come up and shove their hand on or something.

Mary Anne was repulsed by the thought of her stomach being touched by an acquaintance during pregnancy and yet in the final instance she did not protest.

> One of [her husband's name] friends – he's moved now, but – his wife has just had a boy and when he sees me he says: 'How are you? Can I have a touch?' and I think 'Ooo yuck!' The first time I told him 'No, he can wait 'til the baby's born and then touch it' . . . Then the last time we saw him he did anyway, and I thought who cares and I just let him.

In a journal entry Paula explains:

> Sometimes I feel as though being pregnant automatically deprives me of any individual identity and personal space. People seem to have a fascination with pregnant women's stomachs and want to pat them. It's not something they would normally do, but because I've got a 'bump' it seems that I've become public property.

This gazing at, and touching of, pregnant women's stomachs is tied into a notion of the fetus as public property. Wendy Chavkin (1992: 193) makes the point that in the United States a pregnant woman is positioned as 'antagonistic to the fetus if she deviates from medically, socially, or legally sanctioned behaviour'. A number of policies now 'convey a vision of an errant pregnant woman whose antagonism to the fetus must be constrained by outside intervention' (ibid.). Chavkin (ibid.) claims that: 'Autonomy, bodily integrity, and constitutional status are all at stake'. Chavkin's comments help to make sense of Paula's journal entry:

> Complete strangers seem to want to be 'involved' in the pregnancy process. I often get stopped in shops (particularly the supermarket) to be asked when I'm due, how I'm coping with the summer heat etc. Then the advice and personal stories start.

Young (1990b: 160) claims that: 'Pregnancy does not belong to the woman herself. It is a state of the developing fetus, for which the woman is a container'.[5] Does this turning over of the pregnant body to a public gaze and a public touch support Young's claim that the process of 'becoming-mother' is distanced from subjectivity and identity? Perhaps, but not *all* pregnant woman allow their bodies to be touched when that touch is unwelcome and not *all* people consider it their right to touch the stomach of a pregnant woman (which could be read as robbing her of some human subjectivity and agency).

The behaviour of pregnant women is frequently policed not just by health practitioners but also by employers, colleagues, neighbours, friends and loved ones. People frequently regard themselves as societal supervisors of pregnant women's behaviour and so it may make sense to touch a pregnant woman's stomach; to look after that property, that potential citizen in which there is a collective interest. The individual pregnant woman's capacity is primarily as a vessel, while the fetus has a positive and public identity.[6]

Tied in to the idea that the pregnant woman is primarily a vessel for the fetus and that she may be tempted to assert her own primacy (which it is assumed is likely to be antagonistic towards the fetus – see Chavkin 1992) is the idea that the pregnant woman must take care and look after herself (read: take care and look after the *fetus*). There is a tendency to treat pregnant women as being in a 'condition' (see Young 1990b: 170).

One of the ways in which pregnant bodies are constructed as in a 'condition' is through some health professionals continuing to define pregnancy and other reproductive functions as requiring medical treatment. Young (1990b: 168–169 citing Katz Rothman 1979: 27–40) notes:

> even medical writers who explicitly deny that pregnancy is a disease view normal changes associated with pregnancy, such as lowered haemoglobin, water retention, and weight gain, as 'symptoms' requiring 'treatment' as part of the normal process of prenatal care . . . A continued tendency on the part of the medical profession to treat pregnancy and childbirth as dysfunctional conditions derives from the way medicine defines its purpose . . . [that is] as the practice that seeks cures for disease.

This is despite the fact that women often have 'a sense of bodily well-being' and 'increased immunity to common diseases such as colds, flu, etc.' during pregnancy (Young 1990b: 170). This tendency to treat pregnancy as in a 'condition' can lead implicitly to a conceptualisation of women's reproductive processes as disease or infirmity.

The procedures created and adopted by the medical profession inscribe pregnant bodies in complex ways. Today, for example, pregnant women in Hamilton are advised not to smoke or drink alcohol during pregnancy. Most pregnant women will have at least one, probably more, ultrasound scans and most will listen to the heartbeat of their babies through special monitoring equipment. First-time mothers are usually required to attend antenatal classes. Some women will take herbal preparations, such as drinking raspberry leaf tea, for several months prior to the birth, in order to increase the chances of an 'easy' birth. Others may prepare by massaging their perineum with oil, or getting their partners/lovers/friends to massage, in an attempt to make the skin more elastic before delivery so that it can stretch without ripping when the baby's head presses through it. Some women will attempt to rest in the afternoon. Whatever the approach to pregnancy, however, most will be encouraged to see themselves as being in a 'condition' whereby they must take care of themselves and their baby.

The pregnant woman often finds herself being advised continually, not only by medical professionals but also by seemingly well-meaning people such as friends, lovers, husbands, employers, even strangers, to take it easy, don't lift any thing heavy, sit down, be careful, don't stretch, and don't bend.

REBECCA: At work (.) I mean, it's lovely that they care so much . . . but it's like, 'put your feet up, we're gonna get you a foot stool' kind of thing – 'do you want a cup of tea? You're not eating potato chips?' And I'm like, 'I'll eat potato chips if I want to' [laughter] . . . There's a lot of women that lap it up – all the attention. You know, but I'm, I'm like, 'I'm having a baby, now leave me alone and let me get on with having it . . . If my feet are fat [referring to the swelling of her feet due to toxaemia], okay, I'll get off them when *I'm* ready, but I'm not getting off them the minute they go fat.

This comment was reiterated by a number of women involved in the Hamilton research:

DONNA: I wasn't allowed to do any lifting. I got my head snapped off if I even just thought about it.

SAM: I know that from the moment that my grandparents found out that I was pregnant it was almost like, you don't do any activity at all; you sit round with your feet up all the time, you rest and you have afternoon sleeps.

JILL: Mainly, you know, people say to me 'sit down and put your feet up'. I say 'I can't. I want to get this wallpaper stripped.'

It became evident during the research that husbands/male partners, like colleagues, friends, and in-laws, are fully implicated in this discourse of pregnant women as being in a 'condition'.

The position occupied by husbands/male partners, I think, is particularly interesting. Often the fetus represents the joint concern of the 'becoming mother and father' rather than the individual concern of the 'becoming mother'. A linguistic term currently in vogue in New Zealand and in other western countries is 'we're pregnant'. It is now part of the dominant discourse that men ought to 'share' in pregnancy. In a booklet entitled *Your Pregnancy: To Haputanga* (published by the New Zealand Department of Health 1991 and given out to most pregnant women during their first antenatal visit) it is stated: 'Fathers share much of the excitement and worries of pregnancy' (p. 14). It seems that part of this sharing in the pregnancy means offering 'support' to their wives or partners. In another booklet, *Baby on the Way* (1994), which is also distributed to most pregnant women in New Zealand, it is stated: 'While all this [pregnancy] is understandably bewildering for the man, it is important that he understand what is going on and support his partner'. Yet what support entails is not specified.

The word support is defined in the *Collins English Dictionary* (1979: 1460) as '1. to carry the weight of. 2. to bear or withstand (pressure, weight etc.). 3. to provide the necessities of life for (a family, person etc.)'. Support, according to this definition, does not necessarily entail understanding and respect. Bearing this in mind may help to make sense of the following comments made by pregnant women.

A few of the pregnant women talked about their husbands/male partners being 'supportive' but also claimed that their husbands/partners 'growled' at them. Dorothy claimed that her husband 'growled' at her if she mentioned (worried about) how much weight she had gained.

> I've actually put on quite a bit of weight, but he's never really mentioned that, in fact he's always growled at me when I worry about it.

Mary Anne, too, said that she was 'growled at' by her husband.

> Barry did all the heavy lifting, I just did the unpacking and um, moving of the light stuff and um, I moved a couple of heavy things when my husband was working and got growled at when he came home and realised that I'd moved them.

Helen said:

> Gary went ape [became angry] when he found out I'd painted the ceiling. I suppose perhaps I did over do it a bit.

In all these instances, however, the women also described their husbands as caring and supportive. Support and 'growling' were not considered to be mutually exclusive but rather to be mutually constitutive. 'Growling' was interpreted as a gesture of caring and support. These women were quite open to having their husbands discipline their behaviours, in much the same way that an adult might growl at a child. Women are constructed as needing extra guidance, protection and disciplining during pregnancy. They are considered to be prone to behaving irrationally in ways that may harm both themselves and the fetus they are carrying.

Other husbands/partners did not growl but they did offer advice. Paula's husband, Roy, told her to walk (not run) after the ball when they were playing cricket. In another instance, Dorothy said that she was not that keen on going out for walks but her husband insisted that it was a good idea for her to get some exercise.

ROBYN: So does your husband walk with you?

DOROTHY: Mm if it wasn't for him I wouldn't be walking [laughter]. He's the one that makes me go out.

While I did not specifically seek to interview 'becoming fathers', on occasion they were present during interviews. Sometimes they would join in the conversation. Gary, explained that although his partner Helen wanted a home birth he was not keen. His reasons for not wanting her to have a home birth are interesting.

> I am worried about a home birth because again this feeling of responsibility when my wife is pregnant, my role in the whole thing is to have some responsibility towards protecting her and that I have to control things.

When I went to the CBD with Paula, who was at that stage 37 weeks pregnant, her husband Roy accompanied us (or rather, I accompanied Paula and Roy). We spent an hour in a central shopping centre in Hamilton. We visited shops, a bank and had afternoon tea. During this excursion Roy continually stayed very close to Paula. He put his arm around her back, held her hand and guided her by the arm. His behaviour could be read as supportive; it could also be read as protective and/or constraining. Some women claimed that they also received advice from men other than their husbands/partners. One of Kerry's colleagues told her: 'You've got to take it easy. My wife, you know, she overdid it and she was sorry'.

Pregnant women are represented popularly as being in a 'condition' and not suited to the rigours of sport, physical work, 'night-life' and so on. Medical professionals, friends, relations, colleagues and husbands/partners frequently offer 'support' which can serve to disempower and reduce pregnant women's autonomy. Much of this support is given by way of advice – advice about exercising, smoking, alcohol consumption and drug (ab)use. There is also a great deal of advice that focuses on diet and nutrition.

On having pregnancy confirmed by a health professional or during a prenatal first visit to a physician or midwife the 'becoming mother' is sure to receive advice on diet – what the body takes into itself. This advice may be verbal and/or written. In my first visit to the midwife I received verbal advice as well as a number of pamphlets such as 'Listeria in Pregnancy', 'Food Fantastic', and 'Iron in Pregnancy: Nutrition for Two'.[7] While the advice itself has changed over the years, the fact of pregnant women receiving advice on diet is in itself not a new thing.

In nearly every book I have looked at on pregnancy there is a section on diet including books dating back 50 years. For example, Minnie Randell (1945: 23) claims that: 'During pregnancy the prospective mother will be under the care of her doctor or midwife... [and] she will receive instruction in the care of her health and of her diet'. Randell goes on to say that: 'The latter will lie more especially in the selection of suitable foodstuffs which will help to keep the baby small' (ibid.). Citing from a book entitled *Safe Childbirth* by Kathleen Vaughan (no date given), Randell (ibid.) writes: 'The size of the child can be controlled by diet and perhaps more surely by exercise. No child should weigh more than 7 or 8lb at birth'. Randell (ibid. citing Vaughan n.d.) believes:

> Among all nations there are traditional diets for the pregnant woman whose object is to ensure an easy delivery. A vegetarian diet is the most natural one for the pregnant woman – fresh vegetables, fruit, nuts, grain, milk also, but not food out of packets or tins.

Both the universalisation and the naturalisation of pregnancy are evident in this claim about the best diet for pregnant women. Randell (ibid. citing Vaughan n.d.) goes on to say that not only does 'stuffing with oddments, chocolate, extra soup, a little pastry and cake' make the baby enormous but it also destroys the mother's health and appearance.

By the 1950s in New Zealand the dietary advice given to pregnant women had changed. J. Bernard Dawson (1953: 37) claims that: 'It is wrong to have the idea that starvation in the later months of pregnancy will result in an easy confinement on account of the birth of a small baby. The baby will grow at the expense of the mother's tissues if the mother does not take sufficient nourishment'. By the 1950s the vegetarian diet had given way to encouraging pregnant women to eat, amongst other meats, liver (at least once weekly). Also on the menu for pregnant women was one and three quarter pints of milk plus two teaspoonfuls of cod liver oil daily (Deem and Fitzgibbon 1953: 18). The discourse moved from starvation in the 1940s to eating for two in the 1950s. The emphasis on natural food, however, persisted. J. Bernard Dawson (1953: 38) advised:

> With regard to the quality of the food, it is important to remember that fresh natural food is far more valuable in maintaining health and energy in mother and child than artificial or processed food. It is better to obtain one's food from the dairyman, the greengrocer, and the butcher rather than from the chemist.

In the 1960s yet another discourse emerged. This was that pregnant women should not starve themselves in order to produce a small baby which would make for an easier birth, but neither should they 'eat for two'. The idea was that so long as women had a 'sensible' diet there was no need to have extras because the fetus would always take what it needed.

In the 1970s the instructions as to what constituted a 'sensible' diet for pregnant women became more specific. This was prompted by 'scientific research' that suggested 'when pregnant women have an inadequate diet [read: "not sensible"], their babies may die or be born in poor health, and women may have difficult pregnancies and labours, as well as subsequent illness' (Kitzinger 1989: 86).

This idea continued into the 1980s and 1990s. Kitzinger (1989: 86) claims that:

> If a woman is nutritionally deprived her baby is deprived too; she is more likely to have a miscarriage and, if the pregnancy is maintained, the baby is more likely either to be born prematurely or to be of low birthweight because it has not received sufficient nourishment in the uterus. The research also revealed that poor nutrition in the later part of pregnancy can affect the development of the child's brain.

In many of her childbirth manuals, Kitzinger lays out and discusses the nutritional needs of pregnant women in depth. In her well-known book *Pregnancy and Childbirth* (1989: 89) Kitzinger recommends that pregnant women have 92 grams of protein a day for optimum health (this is twice as much as women who are not pregnant). Also, milk is recommended for the pregnant woman, but unless her diet is grossly inadequate in protein she will not need more than 0.5 litre a day. Instructions are also given concerning her intake of carbohydrates, fats, vitamins and minerals. Interspersed with these instructions are many warnings about the dangers of putting on unnecessary weight – 'Cakes, puddings and biscuits do not do much to help your unborn baby's health. If you like sugar in tea and coffee, train yourself to enjoy both of these without it' (ibid.).

I am not, however, suggesting that pregnant women are merely rendered passive and compliant in the dietary regimes prescribed for them. Women often termed 'at nutritional risk' – women who are underweight or overweight, those living on a very restricted range of foods like a macrobiotic diet, regular drug users, women who smoke, and heavy drinkers – may be considered to be resisting hegemonic constructions of the pregnant body in terms of dietary regimes. These women are likely to be read as antagonistic towards their fetus (Chavkin 1992).

Likewise, the pregnant body can also be read as a feminine body that can escape some of the other dietary constraints frequently placed on women such as the need to eat non-fattening foods in order to appear slim. Jude, a university student aged in her late 20s, claims:

> It's nice to escape into pregnancy too, to get away from a slim culture. As a person who has never been particularly pencil thin it is really nice to eat lots of chocolate biscuits and not worry – to be able to hide it with pregnancy. To say 'I'm pregnant, of course I'm big. What do you expect? There's a baby in there'.

To sum up this section, pregnant women have for many years received a great deal of advice on diet and nutrition. This advice has not remained static, but has changed over time (it also varies from place to place and culture to culture, although I have not focused on this). On closer examination, the medical 'facts' of the pregnant body are discourses that change over time and space. Like all bodies, the pregnant body, is subject to specific disciplinary regimes (Foucault 1979, 1980).

To sum up this chapter, pregnant women are constructed as 'modes of seepage' (Grosz 1994a) and as 'matter out of place' (Douglas 1966) in the public realm. Drawing on the notion of abjection (Kristeva 1982) I put forward the idea that pregnant women personify the border between self and other. Their bodies are inscribed as abject. More specifically, they are constructed as seeping (Grosz 1994a) and ugly (Young 1990a) – bodies that do not belong in public space. They are bodies that mark 'sexual difference'. They are bodies whose boundaries are constructed as unpredictable in the public realms. It is unsurprising, therefore, that these pregnant bodies are thought to be in need of surveillance and containment. This desire to keep women contained leads to the fetus/becoming mother being 'taken over' as public property and of public 'concern'. Some of the ways in which this 'concern' manifests itself are that women become subjected to uninvited touch (usually their stomachs), to being in a 'condition' and to specific dietary regimes.

The pregnant body acts as a useful motif for geography's disciplinary body. Both the pregnant body and the disciplinary body possess insecure boundaries. Both are subject to discursive reiteration in an attempt to secure their boundaries. Filtrations and flows of fluids and ideas cannot be stopped. Seepage occurs across the boundaries of both pregnant bodies and bodies of knowledge. Pregnant bodies and bodies of knowledge are spaces of self and other, embodied subjectivity, and politics. In Chapter 6 I extend this argument by considering interuterine space as a close(t) space. This offers a way of rethinking the spatialisation of bodies and the embodiment of spaces.

# 4

# MEN'S BODIES AND BATHROOMS

Much of the large interdisciplinary literature on the body that has emerged over the last decade focuses on women (including menstruating, pregnant and lactating women), lesbians, gay men, 'blacks', the 'poor', aged and/or disabled. These are bodies that are often constructed as Other. Less has been written about the (supposedly hard) bodies of heterosexual,[1] 'white', able-bodied men. As a strategic move to displace the alignment of femininity (homosexuality, 'blackness', disability) with the body and masculinity (heterosexuality, 'whiteness', able-bodiedness) with rationality I conducted research not just on pregnant women but also on heterosexual, 'white', able-bodied men. The aim was to get these men to talk about their bodies, not just as hard, strong and sexualised but as vulnerable and transgressing their boundaries. One route to doing this was to invite heterosexual, 'white', able-bodied men to talk about their experience of domestic toilets/bathrooms.[2]

Toilets/bathrooms are often used as spaces in which bodily boundaries are broken and then made solid again. They are spaces in which bodies are (re)made and (re)sealed ready for public scrutiny. It is impossible to ensure that there are no leakages across the boundaries between inner and outer worlds in toilets/bathrooms. This means that they are often experienced as sites/sights of abjection that are too 'squeamish' (McNee 1984) for the focus of geographers.

The aim of this chapter is to expose the flesh and blood of heterosexual, 'white', able-bodied men so that they can no longer pass themselves off as solid and hard. Theweleit (1987b: 314) discusses some of ways in which young boy's bodies (through the military academy) are reconstructed into soldiers' bodies: the boy becomes 'a man with machine-like periphery, whose interior has lost its meaning'. In toilets/bathrooms it is difficult for men to 'subjugate and repulse what is specifically human with them' (ibid.). By examining the 'toilet/bathroom narratives' of 18 heterosexual, 'white' Anglo/European able-bodied men I hope to peel back men's armour. As has been suggested in Chapter 3, it is often women who become constructed as 'modes of seepage' (Grosz 1994a: 202). Thus, by examining men's corporeal flows I aim to trouble the discourse of women (only) as 'leaky'. Lived experiences of bodily fluids are mediated through cultural representations, images, expectations and place, in this instance, the place of domestic toilets/bathrooms.

## (Not) speaking of bodily fluids

Men's bodily fluids are an underexamined topic. Grosz (1994: 198) explains that when she carried out library research in order to write a book chapter on 'sexed bodies', she:

> was at first puzzled and shocked that where there seems to be huge volume of literature – medical, experiential, cultural – on the specificities of the female body . . . there is virtually nothing – beyond the discourses of medicine and biology – on men's body fluids.

Grosz explains that this has changed slightly with the AIDS crisis but most of this literature remains medicalised.

> There are virtually no phenomenological accounts of men's body fluids, except in the borderline literatures of homosexuality and voyeurism (the writings of de Sade, Genet, and others are as close as we get to a philosophical or reflective account of the lived experiences of the male flow).
>
> (Grosz 1994a: 198)

It was, in part, these comments by Grosz that motivated me to elicit narratives from men about their bodily fluids and flows. Heterosexual, 'white', able-bodied men are often not subjected to rigorous analysis in relation to their embodiment, subjectivity, emotions and private geographies. In academic work on the body men's material bodies are all too often overlooked. Such an omission helps enable white men to retain their position as rational and untainted by the messiness of corporeal flows. I wanted to see if it was possible to get men to talk about their own embodiment, not just the surfaces, but also the depths and flows, of their bodies. I also wanted to see if it was possible to get them to talk to other men about these things.

The men did talk to each other (although not with me facilitating the conversations). Two male research assistants conducted four focus groups with men who live in Hamilton, New Zealand (see Plate 3.1, p. 35 for a location map and photographs of Hamilton). Houses in Hamilton are mainly single-storey detached houses, villas or bungalows. Most of the original one acre rectangular allotments typical of the nineteenth century have now been infilled with other houses (see Porteous 1991) but most New Zealanders still value the privacy that comes with living in detached housing. Early nineteenth century settler cottages did not contain a toilet/bathroom. These facilities were located separate from, and at the back of, cottages.

> In the last two decades of the nineteenth century, however, the bathroom entered the house – a very small room to begin with, just large enough for a tin bath and a basin of tin, enamel, or English porcelain on a wooden stand or case iron brackets. Instead of a bucket to empty the bath there was the novelty of a waste pipe which took the water to a soak hole outside.
> (Salmond 1986: 144)

The toilet did not enter the house until much later. In the 1890s toilets began to be built on to the end of laundry outhouses or at the end of back verandahs of some new houses. 'Others were content to leave it at the farthest corner of the garden' (ibid.).

I began this research on men's experiences of toilets/bathrooms by using my networks of men friends, colleagues and family members. I explained that I was conducting a project on bathrooms and was looking for heterosexual, able-bodied, Pakeha men to be part of a group discussion about toilets/bathrooms. I asked men if they would be interested in taking part. In most cases my explanation and request was met with an outcry of 'no way' followed by laughter and derision about academics and the academy. I decided that a male research assistant might have more success. Perhaps men would be more likely to talk with other men if a man were to facilitate the discussion. It took several months to find a suitable male graduate student who was willing to take on this task. During these months I nearly gave up, deciding instead to write about the fact that so many men had been so unwilling to talk about their bodies, toilets and bathrooms. I was at the point of surrendering to the silence when a male graduate student mentioned that he had some 'mates' who were willing to be involved and that he himself was willing to facilitate the group discussion.

All four of the focus groups lasted approximately one to one and a half hours and were audio-recorded. The 18 men who took part in the research ranged in age from 16 to 58 years (the average was 29 years), worked in a variety of occupations, and lived in a variety of household types (see Table 4.1). The discussion began with the participants sketching their bathroom (see Plate 4.1). This was a good 'ice-breaker'. Following this the men were asked to describe the bathroom they had drawn, including its position in the house, its shower, bath, toilet, and the colour and 'feel' of the room, its size, decor, dirtiness/cleanliness and presence or absence of windows (see Plate 4.2). The participants were also asked how many toilets/bathrooms in their home, whether there was an en-suite, and whether the toilet/bathroom door had a lock that they used. They were also asked to list all the things they do in the toilet/bathroom and whether or not these rooms were places of relaxation. Questions such as: 'how many minutes/hours do you spend in the toilet/bathroom each day?' 'who cleans the toilet(s)/bathroom(s)?', and 'do you examine yourself in the mirror (if your toilet/bathroom has a mirror)?', were also put to the groups.

*Table 4.1* Pakeha, heterosexual, able-bodied men in Hamilton, New Zealand, 1999: profile of participants' general characteristics.

| | *Age* | *Job* | *Shares house with* |
|---|---|---|---|
| *Group 1: tradesmen* | | | |
| Sam | 38 | Lineman | Wife/children |
| Al | 40 | Panel beater | Partner/teenager |
| Ken | 16 | Labourer | Mother/mother's partner |
| *Group 2: colleagues* | | | |
| Brent | 33 | Employment adviser | Wife/child |
| Christopher | 45 | Civil servant | Partner |
| Robert | 24 | Employment adviser | Partner |
| Wayne | 58 | Employment adviser | Wife |
| *Group 3: job seekers* | | | |
| Pete | 25 | Self employed artist/cleaner | Flatmate |
| Tom | 18 | Student | Flatmates |
| Graeme | 32 | Unemployed | Flatmates |
| James | 18 | Unemployed | Partner/flatmate |
| *Group 4: 'mates'/rugby players* | | | |
| Craig | 23 | Bank officer | Flatmates |
| Steve | 25 | Manager | Flatmates |
| Bob | 33 | Furniture company owner | Partner/flatmates |
| John | 21 | Student | Brother/flatmate |
| Matt | 24 | Farmer | Parents/siblings |
| Andrew | 27 | Cabinet maker | Flatmates |
| Alex | 23 | Salesperson | Parents/siblings |

I transcribed each group discussion in full. As I sat in my office hour after hour transcribing this data I experienced a range of emotions. At one point I broke into hysterical laugher causing perplexity and amusement for my colleague in the room next door. At other times I was completely repulsed by the conversation that I was transcribing as the men provided details of their defecation practices. There were also periods of intense boredom where the conversation seemed so banal that I almost forgot why I had chosen this topic. After transcription, I printed copies and read the scripts several times. (See the Appendix for further discussion of the analysis of data.) Several themes emerged from the men's narratives. The first was a fear of (homosexual) contamination by 'circuits of fluid'.

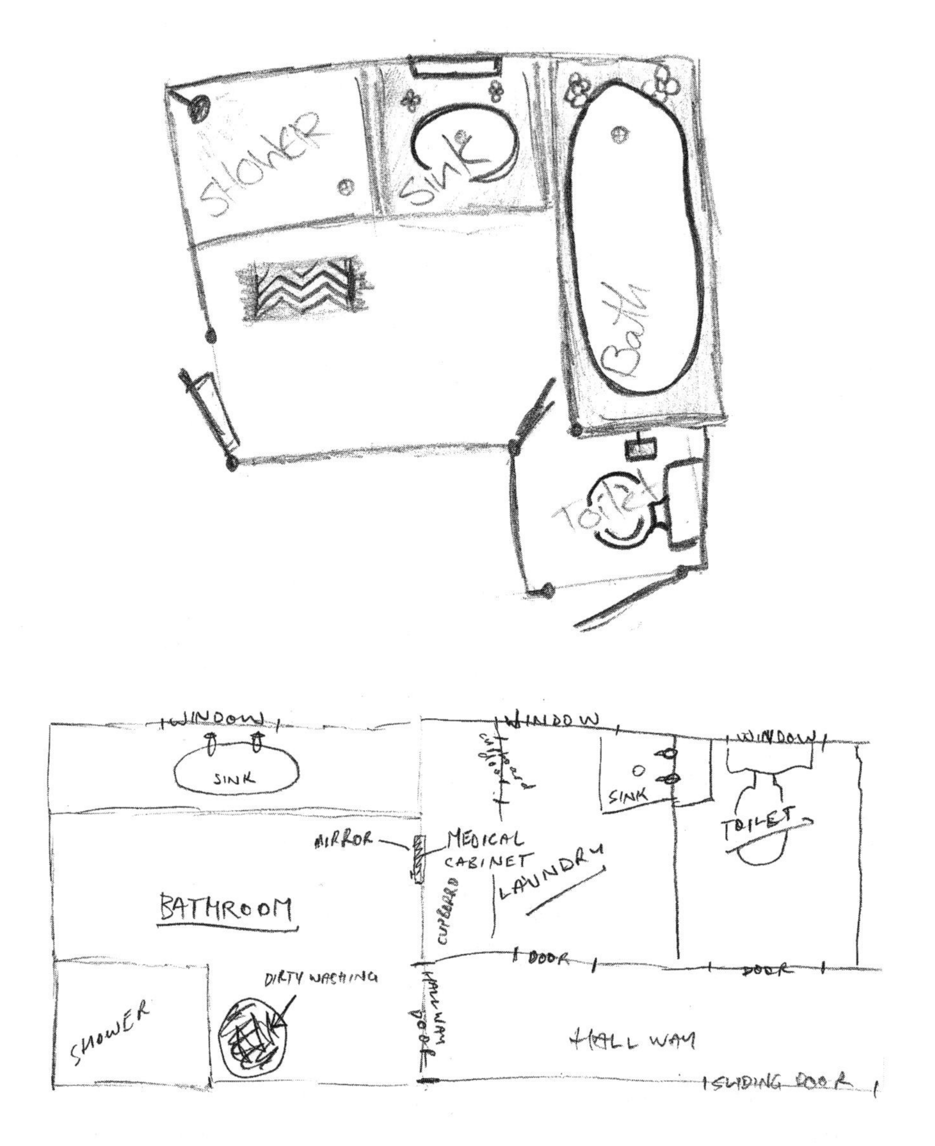


*Plate 4.1* Participants' sketches of their toilets/bathrooms.

*Plate 4.2* A participant's toilet/bathroom.
*Source:* Photograph by Robyn Longhurst, 1999.

## Circuits of fluid

Men who share bathrooms with other men, for example, when 'flatting'[3] often fear being contaminated by other men's seminal fluid. The facilitator asks the men if the shower is an 'appropriate' place to have sex. James, aged 18 and living with his 'girlfriend' and a flatmate, responds:

> I don't know about if you are in a flatting situation. Depends on who is there. Like some people would find it inappropriate. I wouldn't really wanna be having a shower in the same place that your mate and his missus [partner] have just been having sex just before. No, but I don't reckon that it is always inappropriate.

James is not totally opposed to the notion of sex in the shower but neither is he entirely comfortable with the thought of having a shower after his 'mate and his missus' have been showering/having sex.

In another focus group the men are amenable to the idea of their flatmates having sex in the shower but Andrew, a 27-year-old cabinet maker who shares a house with three flatmates, adds the proviso: 'Yeah, it doesn't bother me too much . . . but keep the shower running a little bit after they got out I guess'. The desire to rid the bathroom of seminal fluid after sex is also evident in Bob's (aged 33 and owner of a furniture company) account of 'doing it' with his girlfriend sitting on the vanity. He jokes, saying 'For us five foot eight [inch] guys with short legs it's [the vanity] just the right height' and then adds: 'Only thing is you can tend to get a bit of leakage, runs down the vanity, gotta make sure you wipe it up.'

In these conversations there is a disquiet about seminal fluid, especially if it inadvertently flows between (heterosexual) men. Grosz (1994a) argues that heterosexual men see themselves as an 'active agent in the transmission of flow' but not as a 'passive receptacle'. She continues: 'It may be this, among other things, that distinguishes heterosexual men from many gay men who are prepared not only to send out but also to receive flow' (Grosz 1994a: 201). Grosz explains that it is not the flow in itself that straight men find threatening but that the flow moves in indeterminable directions.

> A body that is permeable, that transmits in a circuit, that opens itself up rather than seals itself off, that is prepared to respond as well as to initiate, that does not revile its masculinity (as the transsexual community does) or virilize it (as a number of gay men, as well as heterosexuals tend to do) would involve a quite radical rethinking of male sexual morphology.
>
> (Grosz 1994a: 200–201)

Some of the men in the focus groups appeared to fear (homosexual) contamination by being placed in a circuit of fluid in their showers (or on the vanity), over which they had no control.

Ken, aged 16, told the story of how he had just spent five hours working as a cobblestone contractor's assistant. He was not used to this work and his hands were red and sore with newly formed scabs over gashed skin from carrying bricks without gloves. The other two members of the focus group, Al and Sam, both in their 40s, suggested that Ken should wear gloves until his hands toughen up. The men also suggested that Ken ought to put something on his hands to toughen them up, for example, methylated spirits. It was also suggested, by Sam, that Ken 'pee [urinate] on his hands'. This suggestion brought general laughter. Al commented that the cane cutters in Australia used to 'pee on their hands' to harden them for field work. Sam retorted (in a jocular tone): 'Then, of course, a lot of them used to die from infection.' To this he added: 'Or perhaps it was a homosexual problem, you pee on my hands and I'll pee on yours'. This comment brought united laughter from the participants.

It is possible to read this slippage from 'peeing on hands' to a 'homosexual problem' in several ways. First, the focus group participants understood the practice of peeing on hands (one's own hands but more particularly the hands of others) to open up a circuit of fluid amongst the men. The shift in discourse from urinating to homosexual acts makes some sense in the light of Grosz's comments about heterosexual men fearing being a recipient of fluid, fearing flow that works in multiple and indeterminate directions. Second, it is possible that the participants understood both 'peeing on hands' and homosexual acts to be dirty – 'matter out of place' (Douglas 1966) – urine not belonging on hands, and gay men and homosexual acts not belonging in the cane cutting fields of Australia. Third, it is possible that the participants conceptualise an unproblematic link between infection and AIDS that they associate with homosexuality. Many readings, some of which are interrelated, are possible of Ken, Al and Sam's narrative.

It was not only seminal fluid (or urinal flow) that appeared to threaten some of the participants. Many were also uneasy about the idea of sharing products that may carry the traces of bodily fluids. Tom explains that he keeps his deodorant in his bedroom because it looks like his flatmate's. He says: 'it looks just the same as his [pointing to his flatmate who was also a participant in the focus group] so I put it in my room. I take it. I don't want wanna be [laughter]'. Tom does not finish his sentence, rather he laughs nervously. It is difficult to determine what Ben does not want to be but his laughter could indicate that he does not want to be intimately connected to his male flatmate through the sharing of bodily fluid (sweat) that might be left as residue on the deodorant.

In this same conversation the facilitator asks these two flatmates whether they take their razors to their bedrooms. Tom responds: 'We just leave them [in the bathroom]. We just leave them sitting on. We've got our own little thing that came with it, you can stand it on.' The conversation revealed that the participants' razors have individual stands that allow the men to demarcate their individual space in the bathroom. The razor stands take away the threat of sharing – the inadvertent transferral of whisker, blood, shaving cream. The facilitator asks Pete, another member of the same focus group, where he keeps his razor.

PETE: My razor, yeah, I keep them separate from everyone else. Deodorant, I keep in my room because, 'cause you are never gonna prove it but you always suspect your flatmates are using it. It really bugs me. Just anyone using my bathroom products really bugs me in general . . . yeah, I think the worst of the worst is if someone uses your toothbrush and you found out . . . I try and keep everything separate aye.

FACILITATOR: So a bunch of stuff would be in your room and you'd bring it through [to the bathroom].

PETE: Yeah

FACILITATOR: And that would be the same with you Graeme?

GRAEME: Only the toilet paper is in the bathroom. You know, in the toilet and everything else is in my room.

The desire to not share bathroom and body products could be motivated by budgetary constraints, that is, the men might not want to share because sharing forces them to have to purchase new products sooner. However, there are indications that more than money is at stake. For example, it is unlikely that a flatmate accidentally using Pete's toothbrush would wear it out but the thought of it causes revulsion.

Another example that more is at stake is found in the interchange between Ken and Al. Ken, aged 16, is financially supported by his mother's de facto partner, Al. The argument about not sharing simply because of the cost of bathroom products does not hold in this instance. During this focus group it transpired that Ken was happy to share with his de facto stepfather, Al, but this did not work in reverse. Ken had not long begun shaving at the time of the focus group. During the discussion Al discovers that Ken is using his shaving cream and that they have inadvertently swapped razors. This occurs when the facilitator asks Ken if he wet shaves or dry shaves.

KEN: Wet, I think? [laughter]

SAM: What do you mean 'you think'?

KEN: Oh, I don't know I just shave.

FACILITATOR: Oh yeah, what do you use?

KEN: Soap, usually, or I've got some shaving cream in a can.

FACILITATOR: What sort of brand?

KEN: Al's usually [laughter]

AL: Mine!

KEN: I used to have a razor, once, I don't know what happened to it. I think Al took it with him when he went to Japan once. Took my razor with him.

AL: *What*? If I did, I left mine behind.

FACILITATOR: What sort of razor? One of those with the blades that (//)

AL: Schick Ultra [a brand name] . . .

FACILITATOR: Is that what you use Ken? I suppose it would be since it's shared [laughter].

AL: It *was* a shared one!

FACILITATOR: It won't be any more! [laughter]

While there is good humour around the sharing of shaving cream and razors it appears that as an adolescent Ken is still to learn the accepted bodily practices and rituals for adult manhood. While Ken is happy to share, Al is not.

Not only did a number of the participants fear contamination through receiving fluid but also they found it difficult to talk of fluid (other than urine), viscous bodily traces or bodily flows. Discourses of bodily fluids and flows were markedly absent in the discussions. For example, there was only one reference to ejaculation (and the word ejaculation was not used – the reference was covert). There were no references to runny faeces, incontinence, bleeding or the expulsion of pus. The men were far more comfortable discussing shaving and excreting.

I suggest this is no coincidence – whiskers and faeces contain the properties of solids rather than liquids. While writing this chapter I came to refer to it informally as the 'geographies of shit chapter' because talk about excrement so dominated the transcripts. Words were used such as 'floaters', 'marble', 'cable', 'shit' and phrases such as 'going for a dump', 'blowing the place up', 'leaving stripes down the side' and 'good long solid one' which indicated that there is a comprehensive vocabulary used to describe the process of defecation. The men (unashamedly) described, in incredible detail, the intricacies of 'doing the paper work', 'wiping your bum', how to flush away the 'floaters' and so on. They described the times of day they defecated, how many times per day or week, and the techniques they used for wiping.

At first I was surprised by their frankness and openness about the biological processes and cultural practices surrounding defecation. The work that I had read on abjection, expulsion and bodily boundary rituals led me to think that these men would deem their excrement to be impure, that it would haunt the edges of their identities and they would attempt to expunge it. This reading was supported in part by some of the participants' comments. For example, Steve and Matt discuss the need for privacy when defecating. Steve concludes by saying that 'going for a dump' is not something that is talked about very much.

STEVE: Yeah, I think it's pretty normal to be secretive about going for a dump. I mean that's pretty trained in you from when you're young.

FACILITATOR: What's the reason though? 'Cause it stinks more or something?

STEVE: It's probably, I don't know.

MATT: It's probably a fair enough point, probably.

ANDREW: I don't think it's 'cause it stinks. It does stink but [laughter].

MATT: Probably because you're doing something that, you know, it's not like you ever go out and go for a dump with someone you know (yeah, yeah) but you can always go out and go swimming naked. Or showering naked. Or doing something naked with someone, but like you never go on a romantic journey and say 'let's go for a dump together'.

STEVE: Yeah, going for a dump is not something you'd ever do in front of someone.

FACILITATOR: You don't have sex when you go to the toilet or something like that do you mean?

STEVE: No. No, not generally [laughter].

MATT: I've never tried that! [laughter]

STEVE: Yeah, it's sort of something that is done just pretty much solid really isn't it? It's never been talked about really has it?

MATT: What?

STEVE: You don't talk about going for a dump much.

The irony implicit in this conversation is that the men were talking publicly and freely about what they claimed was a private act. Their conversation was conducted in complex and nuanced language. It is also worth noting that towards the end of the conversation Steve says: 'it's sort of something that is done just pretty much solid really'. It is difficult to determine exactly what Steve is referring to as 'pretty much solid' but the comment may attest to the fact that the 'mechanics of fluids/solids' Irigaray (1985: 106) suggests underpins systems of signification, logic and symbolisation is apparent – the signification of defecation. It is possible for the men to discuss solids in ways that they are unable to discuss fluids.

A second theme that emerged from the data, and one that is linked to men's disquiet about bodily fluids, is a desire to spend as little time in toilets/bathrooms as possible. Toilets/bathrooms were largely understood as functional spaces rather than as spaces in which to relax and linger. Jeremy Salmond (1986: 201) notes that when New Zealand bathrooms moved inside in the eighteenth century, what was once 'an inconvenient and often uncomfortable chore [bathing] . . . now became a luxury to be enjoyed daily'. For some, however, 'all this preoccupation with the bath and its attendant luxuries indicate[d] the complete weakening of our moral fibre' (Hanna 1931: 30 cited in Salmond 1986: 202). The men in my study did not tend to treat the bathroom 'as a haven of rest and recuperation from the activities of the day' (ibid.).

## Some thoughts on not lingering

Like many of the other participants, Alex, aged 23 and living in his parental home, keeps his body products in his bedroom. Unlike some of the other participants, his decision to do this appears not to be motivated by fear of bodily contamination by others' fluids (although this may be partly the case) but rather by expediency. Alex explains:

> So I just . . . have a shower and then, I've got my deodorant and everything like that, comb and whatever else, in my bedroom so I do that and then, just go to the bathroom and basically just brush my teeth and do my hair.

Alex aims to be in the bathroom for as little time as possible each day. He explains that by keeping his deodorant, comb and 'whatever else' in his bedroom he can be 'in and out'. Alex says he probably only spends 'a maximum of three minutes every morning' in the bathroom.

The participants' desire to not linger in the toilet/bathroom is perhaps surprising in light of a popular discourse that represents men as using toilets/bathrooms as places of escape from domestic tasks – a place to read the newspaper or to do the crossword. Only a few of the participants, however, articulated any desire to spend extended periods in the bathroom relaxing, either while toileting or bathing.

Recent discourses on men's health may have begun to change men's toilet behaviour. Jill Margo a writer on Australian men's health advises: 'Don't dwell on the dunny'. She continues:

> Forget about reading on the dunny. It may be relaxing and it may be a refuge from domesticity or your desk, but it may also be bad for your rectum. The latest medical advice is to be effective at stool. Don't loll about. If you cannot achieve complete satisfaction just clean up and return later.
>
> (Margo 1996: 43)

The focus group conversations, for the most part, reinforced this discourse. Bob explains:

> Like a lot of people, a lot of guys, will go into the bog and have a crap for half an hour, you know, not me, that's just, fuckin', as soon as it's done it is gone.

In fact, men who spend long periods sitting on the toilet were subject to ridicule by some focus group members. Matt, a farmer aged 24, tells the group that one of his previous employers used to go to the toilet every lunch time: 'He'd just sit on the toilet. An hour and a half I swear. I'd have lunch, talk to Sandra for flippin' 20 minutes and he'd still be in the toilet.' Bob replies: 'He's gonna get piles mate. There's no way you can carry on like that.' Andrew, a cabinet maker, aged 27, who lives with flatmates, enters the conversation: 'But I've noticed in our flat there's a chair next to the toilet with magazines on it so obviously the girls like to read a book on the toilet.' Reading magazines on the toilet was something that 'the girls' did, not the men.

In contrast to this conversation, Pete, an artist and part-time cleaner aged 25, explains: 'Sometimes in the morning when I'm home by myself I'll turn the stereo on and take the newspaper in there and have a read.' Note, however, that Pete only does this in complete privacy (although he was prepared to share the information with the group). Christopher, a 45-year-old civil servant, explains that he listens to the stereo and reads when he takes a bath but he does not read on the toilet because it does not feel appropriate and besides he is 'not there long enough'. In response to this comment Robert, a 24-year-old employment adviser who lives with his partner, explains that he also is not there 'physically long enough'. Later in the conversation, however, Robert contradicts himself by saying that he does spend time in the toilet because 'it's the only place I don't get disturbed'.

Most men, however, appeared to spend as little time as possible in the toilet. They did not use it as a place to relax or read. A partial explanation for this is that an anti-intellectual culture exists in New Zealand. Reading is often constructed as a feminised activity. Tom and James explained that reading was not something that they usually do in the bathroom or 'anywhere else for that matter'. Tom says: 'but you are more likely to read a book if you are a reading person. I mean we don't really read books at all do we?' James replies: 'No, I don't read books'. Ken, in another focus group, makes the same point when the facilitator asks him if he reads in the toilet. He replies, 'Not at all. I don't really read full stop'. However, nor did the men report just sitting and thinking, doing puzzles, playing 'gameboys' [brand name of a hand held electronic game] or any other activity on the toilet.

Similar sentiments about not lolling about, relaxing or reading were expressed in relation to bathing and showering. Most men much preferred showering to bathing because it was quicker (most bathrooms in New Zealand contain a bath and a shower – some have showers that are over the bath and some contain a separate shower cubicle). When the facilitator asked the members of one focus group how many of them had baths regularly they responded as follows:

MARK: I don't – no, no.
ALEX: Maybe twice a month.
STEVE: It's been about eight years since I took my last bath.
BOB: I have a bath about once a year probably.
MARK: I have a bath maybe twice a year. Very irregular thing.
FACILITATOR: So the bathroom isn't a place where you go to relax?
BOB: No.
ANDREW: No, nor me.
FACILITATOR: Just in and out sort of thing?

Only a few participants 'admitted' to enjoying bathing and finding it a relaxing experience. Christopher says: 'I like my bath! A bath at night, a shower in the morning'. Brent explains that he sometimes shares a bath with his 18-month-old son: 'he sits on me when we're in the bath together'. Tom and James both claim they enjoy bathing.

TOM: But I don't really use them [baths] for cleaning purposes, more relaxation. Just lying in like a really hot bath.

JAMES: Light some candles or something.

TOM: Yeah, but no so, I mean I agree with what you say but bathing in dirty water, it sort of does defeat the purpose.

FACILITATOR: Could having a bath be seen as relaxation?

JAMES: Oh I suppose so. I mean I've had baths for the sole purpose of relaxation rather than getting clean I suppose.

FACILITATOR: You mentioned the candles. Have you ever done that?

JAMES: Definitely, yeah [laughter]. I only have a bath when I'm tired, or have sore joints, or whatever and I light the candles and it's real nice and listen to some music or something yeah [laughter].

The conversation about lighting candles and relaxing in a hot bath was unusual. Most of the men did not discuss bathing in these terms. When searching the transcripts I noted the absence of words and phrases such as stretch out, soak, bubbles, rubber duck, luxuriate, relax, unwind and bath oils. Nor were bathrooms themselves discussed as sites of home decorating or aesthetic pleasure. Words and phrases such as painting, plumbing, tiling, slate, and mirrors are also absent from the transcripts.

Some of the participants (such as James cited above), however, do mention bathing, or staying in a hot shower for an extended period, when they have sore or tired muscles, usually from engaging in physically strenuous (manly) exercise such as playing rugby.

ANDREW: I personally spend very little time in the shower, I'm not one to, like I've got a flatmate and she will spend half an hour and it just relaxes her and she just loves it. Whereas myself, I just, soon as I get in there I just want to get everything cleaned up as quick as possible and then I'm out. Maybe if I've had a game of rugby or something I'd probably lie there or stand there a little bit longer but, or if you are really cold, but I'm not really a big shower fan though.

ALEX: Yep, I only have a big hot bath after a game of rugby, you might run a big hot bath and have a sit there for half an hour or something.

JOHN: Yeah, mate yeah, in the winter I do, I soak the buns after footy and stuff you know . . . But, yeah just like when it's colder I was sort of brought up with this.

Soaking muscles after a game of rugby seemed to be constructed less as a pleasurable and luxurious experience and more as a way of (re)making the body fit, strong and agile ready for the next game of rugby. In keeping with the attitude that baths were not really places of relaxation or leisure, the men, for the most part, expressed little desire to spend time with their partners, girlfriends or wives in the bath (or shower).

BOB: Baths with women too is just not an option, one of you sort of leans up against the taps and nah, fucking pain in the arse.

ANDREW: It sounds good, yeah sounds all romantic and that but then the bath (//)

BOB: (//) Goes cold!

MATT: Well it depends how sort of young and fresh the relationship is really I think, you know. If is sort of someone that you haven't spent a lot of time with well then you're willing to put in the effort but if it is sort of a long term thing maybe a little bit of the uncomfortability [*sic*] side of into comes into play . . .

STEVE: You've gotta be comfortable all the time. Yeah, yeah . . .

It seems that for these men being comfortable while having sex was a high priority and baths were not considered comfortable spaces.

When the facilitator asked the participants to articulate some possible reasons for showering rather than bathing the responses were varied. Several men mentioned that bathing uses up too much (hot) water. This was a concern for people who were reliant on small water tanks, or tanks that were filled with rain water. Bob explains that he has a bath if:

> my back is rooted [sore] or something and I've got time, but time is a hard thing, to me anyway. The other thing is that I'm on rain water so baths are out because it just takes too much water . . . Baths aren't an option.

Some men also expressed a concern with the electricity costs of heating water for baths, or having insufficient hot water to serve the needs of everyone in the household. Robert comments: 'I've figured out the reason why I don't bath 'cause I used to live in a flat and we didn't have enough hot water.' Similarly Graeme, who shares a house with five others, explains that they have a rule in their house – no baths because they use too much electricity.

Another reason, and one that was outlined by many of the participants, was that bathing is less convenient and more time consuming than showering.

ALEX: I don't have baths because of just the convenience, time.

STEVE: Yeah, yeah

ALEX: You know it takes like ten minutes to run it and you've gotta be in there for quarter of an hour or something like that.

BOB: You always forget it and it'll over fill and you let a bit out.
ALEX: It's sort of a convenience thing, the shower only takes you five, I don't know, eight to ten minutes sort of thing.

This sentiment about convenience ('being quick') applied not just specifically to bathing, showering or toileting but to being in the toilet/bathroom in general. Most of the men seemed to want to spend as little time in bathrooms as possible whether it be shaving, doing their hair or brushing their teeth. When the facilitator asks Steve, manager of a furniture company, how long he spends in the bathroom he replies: 'From going in to starting the shower, about ten minutes'. James, aged 18 and unemployed, also explains that he spends very little time in the bathroom: '30 seconds after every time I go to the toilet . . . shower, I don't know. A few minutes in the shower. Yeah, just washing my face and stuff like that in the morning.' The facilitator asks Ken: 'What about going out at night, you know like going out to a party, you know, Saturday night, would you spend more time in the bathroom before that?' Ken's response is 'Yep', explaining that if he is going out then he does things 'more properly', for example he might put on aftershave and hair 'fudge' (a hair product similar to gel or wax). Al, a panel beater, says that he spends longer in the shower if he is dirty from work.

> Yeah. I find that I've got really dirty fingernails and all that from the job and grease and yeah. And I get sealants stuck all over it and bog [filler] all over it 'cause I finger spread [referring to the sealants and filler used at work]. Yeah it does, it takes a lot longer, and if you're going out, well heaps, it takes longer!

Most of the time though the men appeared to spend very little time in the bathroom. The facilitators asked group members if they had a mirror in the bathroom and whether they tended to use it. Matthew says: 'Nah, I put my hat on to do my hair and it's done'. A discourse of efficiency, of not 'mucking' around in the bathroom, emerged in the discussions. The bathroom did not seem to be a comfortable space for men, or even if it was this ought not to be admitted too readily in the company of other men. There are several possible reasons for this discomfort.

The first reason was put forward by Graeme, a 32-year-old, unemployed man who shares a house with five flatmates. He explained: 'I mean if you spend too long in there people are gonna get suspicious aren't they? So, you don't want to spend too long in there.' My sense is that Graeme is alluding to masturbation – people thinking that if you are in the privacy of the bathroom for 'too long' you might be masturbating. This was one of only two references to masturbation in the 100 pages of transcript material. Both references were covert. Terms such as wank, masturbate or stimulate were not used. Andrew, in a lighthearted tone, explains that he does not take baths because he is tempted to masturbate (although he never uses the word).

ANDREW: I stay away from baths for a simple reason, you're lying in there and you know, you're relaxing and you get a bit bored and give yourself a bit of a wash and then before you know it [laughter].
MATT: Oh what!
ANDREW: And then 'hello' [laughter].

Fear of being understood to be a 'masturbating body', however, only goes part way towards explaining men's reluctance to spend time in bathrooms.

Perhaps a more convincing reason is that bathrooms tend to be coded as a feminine space. This is not unrelated to the argument about masturbation. In talking about running baths, Andrew deploys the feminine pronoun 'she' to describe the bath water. He says: 'You run it, *she's* too stinging hot so you put in a bit more cold and you put your foot in and it's a long process'. Inadvertently Andrew articulates what I had begun to understand during the analysis of the data. Bathrooms were understood by the participants to be places inhabited by bodies that are at the mercy of (Mother) Nature, bodies that are potentially both seductive (including seducing oneself to sexual pleasure) and repulsive. When John tells a story about accidentally 'pissing' into a 'chick's' make-up bag after having a few drinks he refers to his penis as 'she'.

> When I first got up to Aussie [Australia] – the first night I was there I went into two chicks' [women's] place and asked where was the bathroom and I was pissing away, you know, I'd had a couple drinks and then, and yeah *she* started squirting off to the side and I looked down there and I realised that her make-up bag was down there [laughter], all over it. So, I couldn't bring myself to tell them about it so I just had to clean it up the best I could, you know spray a bit of something on it, but Jeez I felt bad though [my emphasis].

The urinating penis that fails to control its own flow is feminised. Other men in the focus group, after listening to John's story, lay the blame for the incident on the owner of the make-up bag.

ANDREW: What is that [the make-up bag] doing sitting down by the toilet?
JOHN: Yeah, that's what I thought.
MATT: It's her own fault mate. It's stupid.
ALEX: It's just common sense .

JOHN: She should shift it when a man comes into the house 'cause it's not like we sit down or anything.

This notion that toilets/bathrooms are a feminised space, or a least they house an insecure, leaky (and therefore feminised) body is also born out in W. H. Auden's poem 'The Geography of the house', written in 1964, about 'a man at stool'.[4] Auden speaks of the inevitability of all bodies ('everybody') needing to defecate. He also alludes to the potential of pleasure (even for melancholics) in this act and the feminisation of Nature.

In short, toilets/bathrooms appear to be threatening spaces for many of the male participants, a space that signals revulsion but also potential desire. They are not places to linger. Metaphors of dirt, filth, shit and crap, and a denial of pleasure, were evident in the men's narratives, not just in relation to toileting, but also in relation to bathing. The participants did not want to be dirty (see McClintock 1995: 209–219 on the relationship between soap, imperialism and the cult of domesticity) but neither did they want to linger (or luxuriate) in toilets/bathrooms.

## Washing in filth

A number of participants argued that not only did they avoid baths because they are time consuming but also because they are 'dirty'. The following conversation was typical.

PETE: I shower, I don't have baths.
FACILITATOR: Why is that?
PETE: You're just basically bathing in dirty water aren't you? Sort of defeats the purpose doesn't it?

Ken reiterates Richard's sentiments:

KEN: No, occasionally I have a bath but not really. Usually I have a shower. Then all the dirt that comes off goes down the plug-hole instead of sitting in the tub with you. I had a bath yesterday after work, I had a bath and the water was dirty as.
AL: So you are the one who left the scum all around the bath then.
KEN: The bath was already scummy. When I ran the water in it was really scummy to start with.

A conversation in another focus group follows similar lines.

MATT: When I first get in there [the bath] I fill it as hot as possible and I get it real nice and warm, like nice and hot and then I'm just lying there and I just start sweating and think shit! What is the point of having a bath? . . .

ALEX: Plus when you're washing yourself you are washing your crap straight into it. With the shower it's recycling the whole time. You know, if you are filthy the bath is the worse thing you can have.

MATT: Yeah, true, that's a good point Alex.

ANDREW: Baths aren't really designed to, maybe in the olden days a bath was designed to clean you but you wouldn't use a bath these days for that.

Even in the 'olden days', however, it is questionable as to how effective baths were for cleaning bodies. These participants add weight to contemporary narratives by reconstructing a history of 'dirty baths' in New Zealand. So too does the facilitator, aged 39, who recalls bathing in his childhood.

> When I grew up in the country and the water that we used to wash in came out of the river-race system, a sort of ditch, and it was full of cockabully [fish] just at a certain time of the year, the settling tank was just filled up with crap. So you'd turn the tap on and this brown shit came out and every now and then a cockabully would pop through and so you'd look at the bath water and you'd just be crazy to get in because you'd end up way dirtier than when you started.

Wayne, aged 58, also remembers baths in his childhood. I quote the conversation at length because Wayne makes explicit the socially constructed nature of bathing and showering. As the oldest member of the focus group he remembers that things have not always been the way that they are now. Showering, Wayne points out, 'is a recent thing'. He explains that being 'smelly' or 'having poor hygiene habits' is less about the individual and more about societal expectations at particular geo-historical junctures.

WAYNE: When I was a kid we had a bath once a week. That is all you were allowed and then it was after the eldest had the bath first and they had that much [indicates about 10cm with his fingers], and to keep it a bit warmer you were allowed to put a bit more in, so the poor little youngest sod at the end almost drowned because the tub was so full. So then they reversed the cycle and had the youngest first in the least amount of water and the oldest moaned because he was sitting in everyone else's shit. And you boiled up the copper and you had to actually bucket it into the bath and so you only had a bath once a week. You were allowed to wash your feet in the (.) and there's only a cold running tap so who was interested in washing their feet under the cold bloody tap in the middle of winter and the bathroom was outside.

FACILITATOR: Where was that Wayne?

WAYNE: New Plymouth, it was in a semi-detached room outside. They had the bathroom and they had the dunny and they had the copper for boiling water in the wash house all in a separate unit, semi covered over, attached to the house. Didn't quite have to walk down the path to the bog [toilet].

FACILITATOR: So you find attitudes from your rural upbringing – [Wayne: this wasn't rural this was in the middle of bloody New Plymouth] provincial city – and what era are you talking about here?

WAYNE: 1940s.

FACILITATOR: So attitudes to bathrooms have changed a great deal since your childhood in NZ? Even when I was a kid it was much more the English thing you know, bathing once a week sort of stuff.

ROBERT: We often bathed more than that but it wasn't every day. And we never had a shower.

FACILITATOR: It's the showering thing, the showering every day thing is quite a recent thing, you think or not?

CHRISTOPHER: I'm not sure.

ROBERT: To me it is because we never had a shower for years.

FACILITATOR: But you've always done it Alan?

WAYNE: Well no he can't of because showers are a recent thing, in the time span of years, showers are very recent things. You could bath every day but not unless you bathed or showered on the beach. You take it as an automatic thing that you do now.

BRENT: I used to bath every day when I started work.

WAYNE: And I reckon a lot of it is society. Society says if you don't bath or shower every day there must be something wrong with you, you must be smelly or you must have poor hygiene habits or whatever.

In this conversation baths are associated not just with cleanliness but also with 'dirt' – people's bodily residue. Wayne explains that the last child to have the bath water was 'sitting in everyone else's shit'.

Baths and toilets/bathrooms, however, were not always understood by the participants to be dirty spaces. At times the toilet/bathroom became more of a (semi-)public space less haunted by the spectre of leaky bodies and at these times it was okay to 'muck around' longer. James comments: 'My girlfriend came in the other day and had a sandwich and talked to me while I was in the shower'. In this instance the often held taboo that it is unhygienic and/or 'improper' to eat in bathrooms is disregarded. This taboo is also disregarded during parties. The facilitator asks the participants if the bathroom, as a space, takes on a different meaning during parties.

TOM: Bathrooms tend to be a real social spot. Yeah.

FACILITATOR: Why is that?

JAMES: Someone will go in there to, I don't know, splash some water on their face and then somebody else will come in for the same reason and go 'Oh Gidday', and sit down.

TOM: Like you tend to think of it as actually not, I suppose not like a, I mean it's clean, but not quite a place you'd hang out but some parties, everything is so busy, so that a lot of people will go to there for a breather or to have a cig[arette]. Yeah, I've actually seen quite a few social bathrooms over the last few years. It just seems to be a place that's not full of heaps of people drinking beer and smoking cigs. For some reason the air seems to be a lot fresher than anywhere else.

In an interesting twist of the more usual coding of bathrooms as dirty places, in this narrative they are recast as clean places – 'a lot fresher than anywhere else'.

## Residue: corpo(real) reminders

Although many of the participants understood toilets/bathrooms to be dirty, few cleaned them. A large literature documents this gender inequality in domestic labour. The narrative below (quoted from a newspaper article entitled 'Sex and filth') sums up the sentiment of much of this literature:

> Beware the humble dunny. If you haven't worked it out already – and if you haven't you're probably a male – it's ground zero in the gender war.
>
> Forget about the old seat-up, seat-down argument, this war of attrition is about who wields the loo brush.
>
> And the sad facts are: in the beginning man created the toilet, but in the end it's almost always the woman who cleans it.
>
> (Dixon 1999: J5)

When the facilitator asks: 'Who cleans the toilet and bathroom?' Sam responds:

SAM: Anne [his wife], that's her job. But I don't want to make too much of a mess in the bathroom because of the fuss women tend to make.

AL: They get really upset about it.

SAM: I'm not the tidiest of people.

AL: Women tend to use bathmats and things and worry about whether toilet seats are left up or down.

Pete, however, lives with just one male flatmate and they share the cleaning of the bathroom.

PETE: We don't have any roster for that you see, and my flatmate he is quite, he is very clean so he tends to clean it systematically, like he'll clean it every week. But with me, I just get these impulses so I might clean it, you know, I might clean it one day, then I might wait a few days and then I'll just get this impulse to clean it again. I never actually set up a time or any thing to do it. I don't know, I just get these gut reactions to say clean it. You know.

FACILITATOR: What about the toilet?

PETE: Yeah, same thing. Normally if you've just used it, and you've really blown the place up I'll clean it then as well 'cause of the smell, so you pour on heaps of chemicals and you think I'll just pour on heaps of chemicals to make the whole thing smell nice and I'll just keep cleaning it, so, yeah.

Pete's comment about the need/desire to clean the toilet if 'you've really blown the place up' [splattered excrement] was reiterated several times in discussions. Given that most of the men claimed that they did not clean the toilet/bathroom it seemed somewhat paradoxical that they would want to clean any obvious 'remains', such as 'stripes', left in the toilet bowl. The low water level in New Zealand toilets (unlike many United States and United Kingdom toilets) means that faecal matter sometimes 'stripes' the inside wall of the toilet bowl. These marks often remain after flushing. Most of the men attempted to clean these stripes before leaving the bathroom.

FACILITATOR: In your house if there's a whole lot of crap left in the toilet bowl, does that become an issue of different expectations about how clean the toilet bowl is going to be?

AL: Well I don't know, I always look, when you've done it and if it's a mess well you clean it up yourself before you go. You don't want someone else coming in and seeing it. It just depends like sometimes the toilet doesn't flush it all away, for one reason or another, 'cause someone's just been there and there's not a full cistern, well you just sort of come back and do it in a few minutes.

Al's motivation to come back and flush the toilet a few minutes later appears contradictory given that earlier he admits: 'I don't help' even though 'my partner works full time'. Andrew also states that he does not clean the bathroom or toilet, unless he leaves 'stripes':

ANDREW: Unless you sit down you know and you leave a few stripes down the side of the bowl [laughter].

BOB: I leave those for somebody else.

ANDREW: Oh no you can't do that!

Andrew is surprised that Bob leaves these 'stripes' for somebody else to clean.

While there was some disagreement about whether it was acceptable to leave 'stripes' in one's own toilet, most were in agreement that it is embarrassing to leave 'stripes' or 'floaters' (excrement that will not flush) in somebody else's toilet. Matt offers advice.

MATT: The trick is, like if you visit somebody else's bathroom and you don't want to do that, you always stick the paper in first before you have a shit.
ANDREW: Oh yeah, no.
ALEX: What?
MATT: Stops the splash, everything.
ALEX: What do you mean put the paper in first?
MATT: Put the paper in first, have a shit,
[talking over each other]
BOB: If you've got a marble and it goes plop, splash.
ALEX: No, I've never done that.
ANDREW: It's crucial, you always put the paper in first. If you get a bit of a cable and she [sic] swings around to the left on you, oh yeah, hits the side of the bowl.
FACILITATOR: . . . So you are less concerned about what happens at your own house, more at somebody else's house? Like if you are at your own house do you?
STEVE: I always use the toilet brush at my own house if I've made some stripes as Andrew called it because you know four people are using the toilet.
FACILITATOR: Is that standard for you guys? If you are at your own house you clean it up?
ALEX: I mean if you made a mess on the seat you'd clean it up at your own house.
MATT: I'd never clean it up at my house.
ALEX: I just always clean it up anyway. I mean if you are at someone else's house you clean it up but you don't go scrubbing the bowl or anything like that, just in and out you know . . .
BOB: The bastards are the ones that you bloody flush the bog and then it comes, boom, back up again, yeah, fucking floaters.
MATT: Polystyrene!
BOB: You've gotta stand there and wait for the bloody system to fill up and give it another go and she [sic] just about gets down and runs out of omph [sic] and goes 'boom' [laughter]. Fuck, I just leave it there [laughter]. Yeah, no good if you are at somebody else's place.

An ambiguity exists here. Most of the men do not take responsibility for cleaning their bathrooms/toilets as such but they do not like their bodily residue (referred to here by Andrew and Bob as 'she') to be visible to others. There is a squeamishness and vulnerability associated with others viewing that which is expelled from inside the body. But there is a contradiction here.

In one of the conversations some of the men almost bragged about the faecal matter they produced. Runny stools may not have been worthy of pride but 'long solid' stools, or stools bearing traces of liver or dark beer were a triumph.

Andrew: How many of you guys if you're having a shit and it's a good long solid one you want to fuckin' look at it aye? It's good – cable!
Alex: You can look at it after it's in the bowl.
Matt: You should have some liver mate. If you want to have a good shit go eat some liver and away to the toilet you go.
Andrew: What about Guiness [beer]? What does that do to it?
Matt: Molasses!

In this conversation vulnerability becomes (re)constructed as its binary opposite, powerfulness (see Campbell 1997 for a critical reading of the term 'vulnerability' in geographical discourse). Feelings of impotence, forcelessness and emasculation over broken corporeal boundaries and bodily residue are recast as an attempt to appear powerful, forceful and robust. In this instance 'shit' is turned into a positive signifier.

## Why bother with the banal?

> Excrement poses a threat to the center – to life, to the proper, the clean – not from within but from its outermost margin. While there is no escape from excrementality, from mortality, from the corpse, these do not or need not impinge on the everyday operations of the subject or body. The (social and psychical) goal is to establish as great a separation as possible from the excremental, to get rid of it quickly, to clean up after the mess.
>
> (Grosz 1994a: 207)

Men are taught to understand themselves as hard, solid bodies. Bodies, however, are also soft, viscous and liquid. They absorb and excrete flow. Domestic toilets/bathrooms are often the site of this flow. Men cannot deny or escape their excrementality, their liquidity, their humanity, in toilets/bathrooms.

Domestic toilets/bathrooms have not simply been excluded from geographical discourse because they are banal or unimportant (they are as 'political' as any other space) but because they are threatening. They are most threatening to those who play a role in constructing what counts as 'legitimate' knowledge in geography and who can bear that knowledge. Although the body has become far more evident in the geographical literature over the last few years there are still some abject sights/sites that remain unspeakable. These sights/sites threaten to spill, soil and mess up, clean, hard, masculinist Geography. There are limits as to what we can do and say in geography. We may be able to discuss discursive constructions of embodiment but we still cannot talk easily about the weighty, messy materiality of flesh, or the fluids that cross bodily boundaries on a daily basis.

One of the reasons for this dis-ease, this dis-comfort, over what are considered to be abject sights/sites is the privileging of the mind over the body in geographical work (Longhurst 1997: 486–501). Adding the usually private spaces of toilets/bathrooms and bodies that break their boundaries to the public discourses of geographical knowledge challenges with the epistemological masculinist underpinnings of geographical knowledge.

Theories of abjection and the destabilisation of binary discourses of inside/outside of bodies are useful not only in understanding pregnant bodies but also heterosexual, 'white', able-bodied men's bodies. In this chapter I have attempted to scratch the surface of Geography's 'hard' crust. Such scratches, cracks and interstices create small openings through which to slip Otherness into the discipline. That which is coded by masculinist hegemony as abject, embodied and feminine cannot always be expelled from geography. Despite discursive reiterations, geography's boundaries are insecure. Its corpus leaks and seeps and in this I take pleasure since it signals future possibilities for contestatory, and potentially, emancipatory, geographies of difference – in this case, a geography of toilets/bathrooms and men's bodily fluids.

In the next chapter I continue to pursue ideas about boundaries, fluidity and leakage but shift the focus from the private spaces of the bathroom to the public spaces of the workplace – shop floors, offices, boardrooms, and inner city shops and streets. I examine the ways in which men and women managers manage their own bodies and the bodies of their employees (the corporate body).

# 5

# MANAGING MANAGERIAL BODIES

> Recently, Kay, a friend in her late 30s, invited me to go shopping to assist with the purchase of a 'corporate suit'. She was close to completing a university degree and was being interviewed for a professional position. In an attempt to persuade me to go shopping, Kay explained that as a five foot, two inch, 'large' woman she was concerned that her prospective employers might read her body as 'out of control'. Kay felt that she had to have a 'smarter' suit than any men (or slimmer women) being interviewed for the position. We shopped for a suit and finally purchased a (very expensive) skirt and jacket with straight, long, 'clean' lines. It was cut from a dark (but not black), heavy (but not too heavy) fabric. Kay was delighted with the way that the suit transformed her 'short', soft, feminine looking body into a longer, firmer, more masculine looking, body. Kay retained some of the markers of femininity (skirt not trousers, sheer tights, light blouse) but adopted some of the markers of masculinity (formal, dark jacket, long, firm lines) in an attempt to ensure just the 'right look' for securing the position.

This vignette about Kay's desire to purchase the 'right' corporate suit for her interview points to some of the (often unwritten) rules about bodily conduct and presentation in 'professional', 'respectable' workplaces. Respectability, argues Young (1990a: 136):

> consists in conforming to norms that repress sexuality, bodily functions, and emotional expression. It is linked to an idea of order: the respectable person is chaste, modest, does not express lustful desires, passion, spontaneity, or exuberance, is frugal, clean, gently spoken and well mannered.

Large-bodied people (especially if they are 'short') are often read as lacking in respectability. They are thought to be unruly, unclean, loud and poorly mannered. Kay was attempting to lengthen, slim down and firm up the lines of her body not only in order to guard against the transgression of physical bodily boundaries but also the social boundaries of respectability.

The focus of this chapter is the bodies of managers in Auckland, New Zealand and Edinburgh, Scotland. I examine the clothes they wear, their health, levels of fitness, grooming and comportment. There are normative expectations that managers will 'do' work in specific ways. Managers are expected to act in a manner that is respectable, professional, rational, in control and well groomed. Young (1990a: 139) notes that: 'The norms of "professional" comportment entail repression of the body's physicality and excessiveness. It goes without saying that respectable norms require eliminating or covering all bodily odours, being clean and "clean-cut"'. Such behaviours are repeated over time and space, eventually taking on the appearance of the 'natural' (see Butler's 1990 discussion of performativity and repeated stylised acts taking on the appearance of the 'natural').

In June 1997 I began to collect data on managerial culture. I read business magazines, advertisements, cartoons, newspaper articles and academic literature. I focused mainly on middle managers because they are often involved with their customers, staff, senior managers and owners on a daily basis. Middle managers not only manage their own bodies but also the bodies of their workers.

I decided to choose a city that was reasonably 'large' (with a population of at least half a million) because I wanted to examine the formality of professional dress codes and conduct in densely populated central business districts (CBDs). In smaller cities these codes are often assumed to be rather more relaxed. Auckland offered a research field in which I felt reasonably familiar and comfortable. I live in Hamilton, a city located one and a half hours' drive south of Auckland. Metropolitan Auckland is easily accessible and thus feels like an extension of 'home' – Hamilton. For many years I have visited Auckland once every few months. Auckland provided a useful place in which to examine 'managerial culture'.

Shortly after collecting data in Auckland I undertook study leave in Edinburgh. This provided an opportunity to collect more data. Edinburgh offered a research field in which I felt 'displaced' (see Katz 1992 and 1994 on 'questions of fieldwork in geography' and on ethnographers as displaced persons). I had only visited Edinburgh once briefly before conducting this research while on study leave. I understood little of 'Scottish culture' and hoped that my 'displacement' might enable me to gain a more nuanced understanding of normative bodily practices for managers. My hope was that in examining Auckland and Edinburgh both similarities and differences would emerge.

I interviewed 26 managers in the CBDs of Auckland and Edinburgh who were employed in a range of industries including travel, food, information technology, security, retail, accommodation and leisure. I was interested in the way that particular socio-cultural environments construct bodies with particular capacities and desires, therefore, I bounded the study spatially rather than examining one particular industry.

In the CBD of Auckland (mainly in central Queen Street) I conducted interviews with 14 managers at their places of work (see Table 5.1). Auckland is by far New Zealand's largest city. In 1996 its population was recorded at 1,077,205 (Statistics New Zealand 1996: 21). The area defined by Statistics New Zealand as Auckland

*Table 5.1* Managers from a range of industries working in the CBD Auckland, New Zealand, 1997: profile of participants' general characteristics.

| | *Age* | *Ethnicity (self-defined)* | *Manager of . . .* | *Salary package for 1997* | *Approx. hours worked per week* |
|---|---|---|---|---|---|
| Debbie | 35–40 | NZ European | Hotel/casino (human resources) | NZ$85k | 70 |
| Gillian | 35–40 | British | Toiletries/cosmetics store | unknown | 50 |
| Yvonne | 25–29 | NZ European | Health club A | NZ$38k | 45 |
| Colin | 45–49 | NZ European | Bank (commercial division) | NZ$80–100k | 50 |
| Trevor | 30–34 | NZ European | Security firm | NZ$100k | 35 |
| Robert | 50–54 | NZ European | Employment consultancy | NZ$80k | 60 |
| Pippa | 25–29 | NZ European | Travel | NZ$35k | 50 |
| Lex | 30–34 | Pakeha | Health club B | NZ$42k | 45–60 |
| Scott | 30–34 | NZ European | Insurance office | unknown | 60 |
| Sandra | 25–29 | European | Café | NZ$60k | 45 |
| Guy | 30–34 | Pakeha | Computer programming company | NZ$60–80k | 60–100 |
| Harry | 30–34 | NZ European | Outdoor recreation retail | NZ$42k | 45–50 |
| Jerome | 30–34 | NZ European | Bank | NZ$65–70k | 45 |
| Hilary | 30–34 | NZ European | Clothing retail | NZ$50k | 50 |

Central experienced a 12.2 per cent growth rate between 1991 and 1996 (ibid.). The area in which I interviewed managers experienced a huge construction boom in the 1980s, much of it in the form of new office buildings. Following the 1987 sharemarket crash, however, the level of building activity dropped dramatically. Today the CBD is dominated by tall buildings (approximately 25–30 storeys high) including the recently completed 'Sky Tower' which contains a hotel, shops, restaurants, cafés and bars, casino and enclosed viewing decks (see Plate 5.1).

Auckland is made up of a variety of ethnic groups, the three most dominant being Pakeha, Pacific Island and New Zealand Maori (McKinnon with Bradley and Kirkpatrick 1997: 94). Both in the past and in recent years Auckland has also experienced migration from various parts of Asia. The managers with whom I spoke, however, were all Pakeha. The CBD is dominated by Pakeha. I was interested in examining hegemony and normative bodily performance, therefore, Pakeha were a useful group to interview. Over half of the interviewees were aged between 30 and 34. The remainder ranged from their late 20s to early 50s. Eight participants were men and six were women. In addition to interviewing managers I also observed their places of work. Sitting in shops and offices, listening to conversation, having an opportunity to examine managers' attire, bodily comportment and grooming was a significant part of the study.

I repeated this data gathering process in the CBD of Edinburgh (mainly in Princes Street and George Street) from December 1997 to January 1998. Edinburgh is the capital city of Scotland (see Plate 5.2). In 1993 the Edinburgh District's population was recorded at 441,620. The 1981 and 1991 censuses indicate that Edinburgh experienced 56 per cent growth in the banking and finance sector between 1981 and 1991. Business services grew by 91 per cent (Edinburgh Facts and Figures 1995 online – 26 November 1999). I interviewed 12 middle managers in Edinburgh who worked in a range of industries similar to the Auckland group (see Table 5.2). The age of the Edinburgh participants varied between 20 and 49. One participant defined herself as Irish, another as Irish/Scottish and another as Scottish/white. The remaining nine participants defined themselves as a combination of British and white. One participant also added the term European to describe his identity.

The interviews in both Auckland and Edinburgh lasted, on average, 30 minutes (in general managers were too busy to take more time than this out of their day). The interviews were semi-structured in that I followed a series of questions but also allowed the interviewees to take the conversation in directions that they thought relevant and interesting. Topics covered included health, exercise regimes, work related illness, stress and stress relief, number of hours worked per week, travel to and from work, dressing for work, and the significance of gender in carrying out their duties as managers. All the interviews were audio-taped and transcribed in full. In addition to being interviewed, all participants, with the exception of one, filled out a two-page questionnaire providing information about their age, gender,

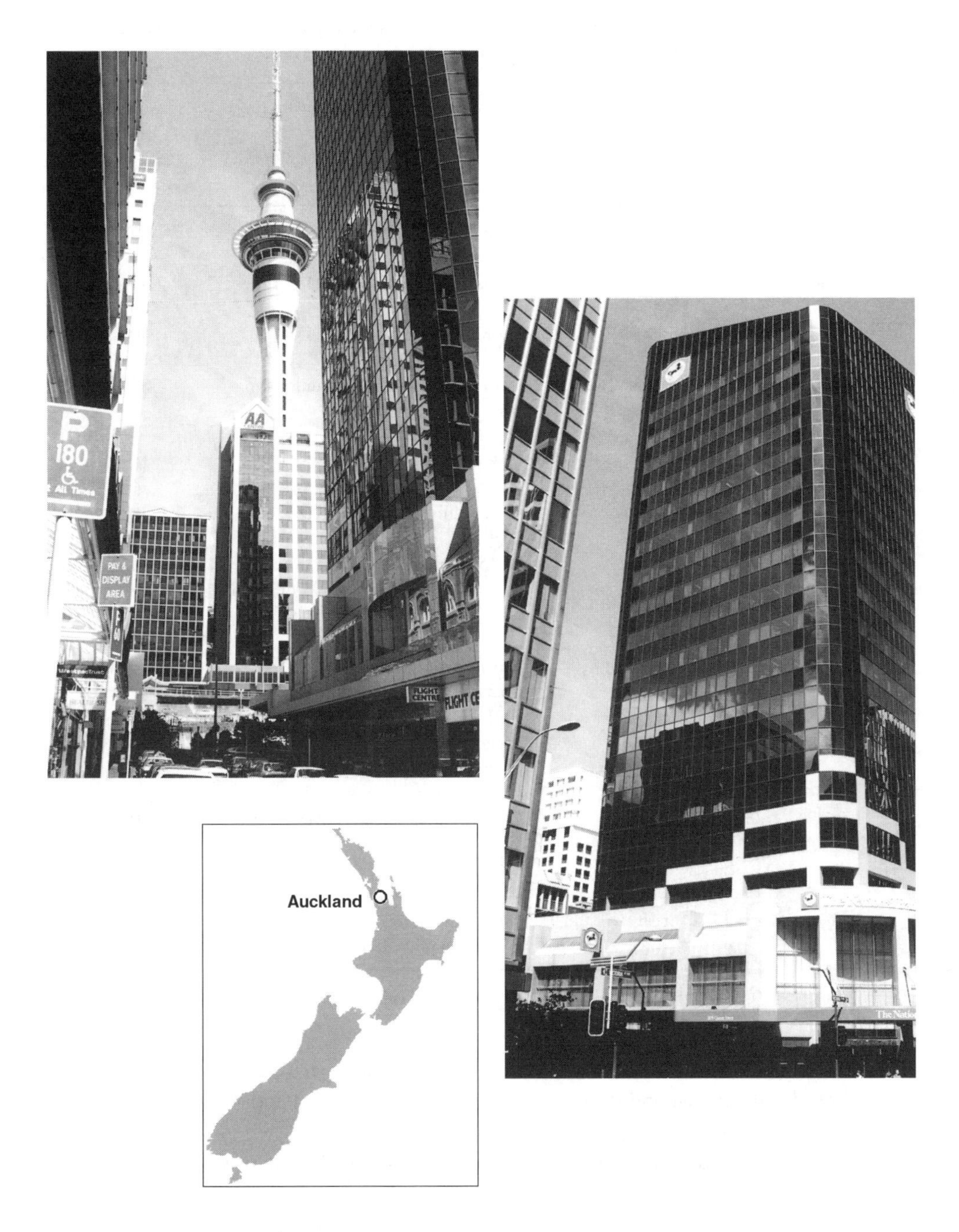


*Plate 5.1* Auckland CBD – photomontage of interview sites.
*Source:* Photographs by Robyn Longhurst, compiled by Max Oulton, 1999.

*Table 5.2* Managers from a range of industries working in the CBD Edinburgh, Scotland, 1997–1998: profile of participants' general characteristics

| | *Age* | *Ethnicity (self-defined)* | *Manager of . . .* | *Salary package for 1997* | *Approx. hours worked per week* |
|---|---|---|---|---|---|
| Bryan | 20–24 | British | Travel | unknown | 37.5 |
| Astrid | 35–40 | Scottish/white | Furnishings store | UK£12k | 45 |
| Charles | 45–49 | British | Clothing store | UK£14k | 40 |
| Ruth | 20–24 | British/white | Women's fashion store | unknown | 37 |
| Peggy | 35–40 | British | Bank (mortgages) | UK£22 | 40 |
| Liz | 30–34 | British | Health club | UK£19k | 50 |
| Dick | 25–29 | White European | Bar/restaurant | UK£15k + accommodation + bonuses | 55–60 |
| Kathryn | 35–40 | White | Department store A (financial division) | UK£32k | 40–42 |
| Wendy | 20–24 | White | Department store A (trainee manager) | UK£16k | 40 |
| Nigel | 25–29 | White | Department store B | UK£20k | 50 |
| Mike | 30–34 | Irish/Scottish | Fashion store | unknown | 50–60 |
| Lyn | 25–29 | Irish | Hotel (personnel and training division) | UK£20k | 60–70 |

ethnicity, job title, daily activities at work, annual salary package, hours worked per week, health status, and dress codes at work. (See the Appendix for more information on data collection and analysis.)

In analysing the interview data two themes emerged. The first was presentation at work – clothing, image, style and grooming. Managers discussed both their own presentation and the presentation of their workers. In particular both Auckland and Edinburgh interviewees paid a great deal of attention to the need to wear a business suit. The second theme was the desire for a particular kind of material body – a

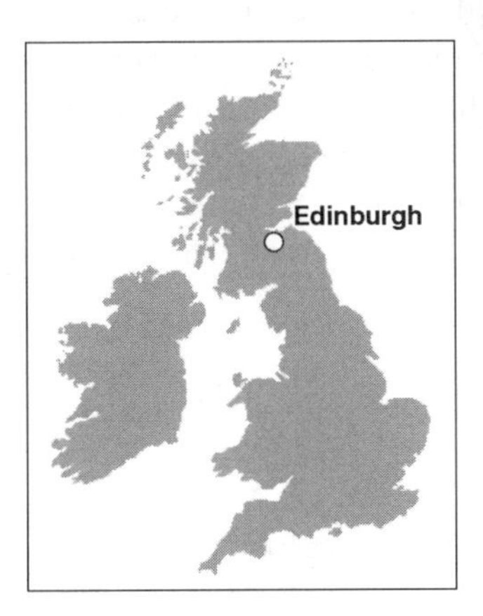


*Plate 5.2* Edinburgh CDB – photomontage of interview sites.
*Source:* Photographs by Robyn Longhurst, compiled by Max Oulton, 1999.

body with a particular weight, shape, size and comportment. Managers, especially in Auckland, appeared to seek high levels of fitness and flexibility. They were intent on crafting a corporeal self that was trim, taut and looked terrific (see Bordo 1990 on 'reading the slender body'). I address each of these themes in turn.

## Suiting the body

Most interviewees in Auckland and Edinburgh dressed in 'corporate uniform' or 'a business suit' (trousers, jacket, shirt and tie for men and trousers or skirt, blouse and jacket for women). In particular, managers in banking, insurance, travel and the hotel industry routinely wore business suits. Jerome, an Auckland bank manager in his early 30s, explains: 'The bank does have a corporate wardrobe . . . For management I guess, a suit and collar, a tie is required, and for ladies the *normal sort of corporate wear*' (my emphasis). Jerome's comment reiterates the notion that corporate wear is normative. Mark, manager of an Auckland insurance company, comments: 'The expectation is that you wear a suit and tie, and when I say "suit", it doesn't mean a sports jacket and tie, I mean a suit. Yeah, that is the expectation'. Lyn, manager of a large and prestigious Edinburgh hotel, claims: 'Managers don't tend to wear a uniform, they wear smart sort of clothing and . . . we've got the name badge. You've got to wear the name badge all the time.' Kathryn, in her late 30s, is a manager of a large department store in Edinburgh. She compares her business attire with that of her male colleagues:

> My male colleagues are expected to wear a suit. A dress jacket and trousers would probably be all right in some stores. It really does depend on the manager of the store but definitely a suit for men. Women are a bit more versatile but I personally would never wear anything but a jacket. So maybe a dress and a jacket, trousers and a jacket, a shirt and a jacket. I would always look as though I had a suit on. And I think that is fair to say for most management.

Nigel, also a manager of a large department store in Edinburgh, details the dress codes operable in his workplace:

> We have a management dress code which we have to follow . . . so that we're not coming to work wearing bright orange shirts or razzamatazz ties and things like that – they are not a good idea. There's certain colour codes that you have to follow – mainly white, well, it creates a *business edge*. White, blue, cream, really . . . Your bright blues and bright green shirts and things like that are not acceptable. It's really looking for the business look. That's what we're looking for [my emphasis].

Nigel's use of the phrase 'business edge' can be read in several ways. First, it can be read to refer to the company needing to secure and maintain a competitive market position. Second, it can be read to refer to creating a body line (edge) that looks hard and impenetrable.

Comments such as Nigel's beg the question: why is 'the suit' or 'corporate uniform' so important for managers? I suggest that it is because the highly tailored, dark-coloured (often black, dark grey or navy) business suits function to seal the bodies of men and women managers. Firm, straight lines and starched creases give the appearance of a body that is impervious to outside penetration. Outside penetration could include the penetration of the mouth with food. Valentine (1998) discusses the 'production of the civilised street', arguing that some people today 'snatch' a bite to eat as they walk down the street at lunch or break times, but others understand this to be 'inappropriate' behaviour. To be seen eating in public is to be seen breaking one's bodily boundaries.

Business suits not only give the impression of a body that is impervious to outside penetration but also of a body that is impervious to the dangers and threats of matter that is inside the body making its way to the outside. It is considered inappropriate for matter to make its way from the inside to the outside of bodies (for example, farting, burping, urinating, spitting, dribbling, sneezing, coughing, having a 'runny nose', crying and sweating) in most inner city workplaces. This observation about people wearing business suits avoiding farting, burping and so on in public almost seems too banal to mention. The performative (Butler 1997) takes on the appearance of the natural. Phillip Garner (1983: 30) destabilises this naturalness in his comic photograph of the 'half-suit' (see Plate 5.3).[1] He claims that the 'half-suit retains the crisp formality established in the neck/collar/tie/lapel area but offers an *abbreviated midsection* for comfort and physical flair' (Garner 1983: 31; emphasis in original). It is the absurdity that makes the photograph work as comic.

The hard lines of the business suit are aimed at making the 'volatile body' (Grosz 1994a) that continually threatens to leak and break its boundaries appear more solid. Peggy, an Edinburgh bank manager, claims that: 'Old school ties, public schools are still really, really there in Edinburgh so you have to be quite hard . . . full of yourself'. The suit helps to create an illusion of a hard, or at least a firm and 'proper', body that is autonomous, in control, rational and masculine. It gives the impression that bodily boundaries continually remain intact and reduce potential embarrassment caused by any kind of leakage. When bodies are draped in soft, light fabrics it is often possible to see the boundaries of the body – the rise and fall of the chest, mound of the breast, contour of the muscle. It is possible to see a spot of blood, a smear of dirt, a piece of flesh. Such matter signifies a body that cannot be neatly contained, a body that is not always rational and in control, a body that is both desirable and disgusting.

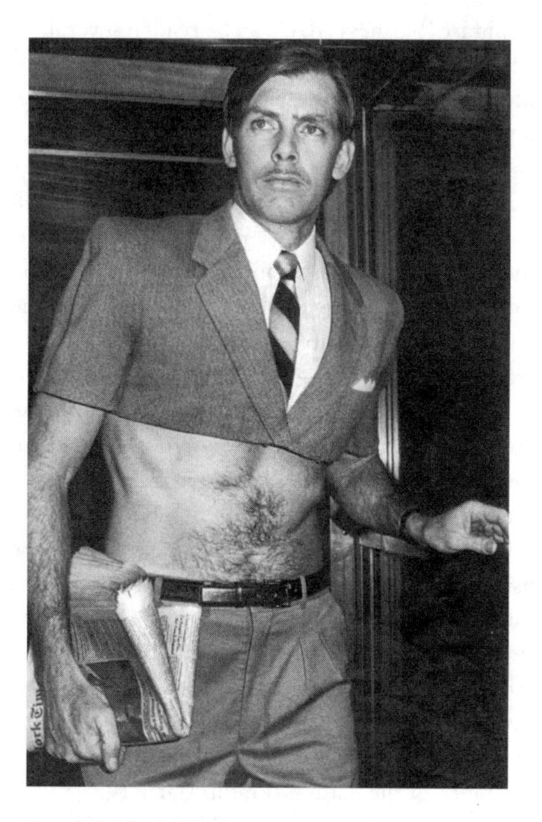

*Plate 5.3* The half-suit.
*Source:* Garner 1983: 30 – photograph by James Hamilton.

It is also considered inappropriate for the professional body in the CBD workplace to take matter into itself (for example, swallowing tablets, sucking objects such as fingers, hair or pencils, using a nasal spray, placing a finger or object in a bodily orifice). Eating, drinking and smoking are permitted in some workplaces but not in most. For example, most managers would not eat, drink or smoke when they are with clients, on the 'shop floor' or during an interview with a researcher. If these unwritten social codes of bodily behaviour are not followed then the individual concerned risks being ridiculed or even losing his/her job. Young (1990a: 137) notes that 'Speech is also governed by rules of decency: some words are clean and respectable, others dirty, and many, especially those relating to the body or sexuality, should not be mentioned in respectable company'.

Some interviewees expressed the opinion that it was more important for the body to appear 'solid', rational, respectable and professional in the city rather than in the suburbs. Lois, manager of an Auckland travel consultancy, explains:

> I think that in this office, because we're in the city and we're dealing with businessmen and women, that we do try to be perhaps a little more professionally dressed and professional in outlook than perhaps if you're in the suburbs where you're having, you know, housewives coming in in their track-suits . . . We don't get many retired people, many housewives, anything like that.

Lois assumes that travel consultants who work in suburban offices serving the needs of (track-suited) housewives and retired people do not need to dress as professionally, or be as professional in their outlook, as city-based consultants. Robert, manager of an employment consultancy firm, also points to differences between the city and the suburbs in Auckland. He says that while having a Queen Street address is advantageous in terms of the status it carries, one of the disadvantages is that it can prove both expensive and alienating for some clients:

> On the down side, for a lot of people coming in to see us it's a real pain because of parking and access problems. I mean, for a lot of 'temps' who are perhaps the lower end of the income spectrum, you know for them to come in and park in down town, walk up here for a three-hour registration and the testing process to register as a temp. I mean you're looking at $30 . . . And a walk. And if they're from the country – for many of them it's, you know, it's pretty glitzy Queen Street . . . this is where you see a lot of people wearing, guys wearing suits and women wearing expensive business clothes and if they're trooping in in their jeans and their sneakers because they're a process worker or a driver or a nurse aide then this isn't their territory.

Managers are clearly delimited from Others – those who do not carry out professional work in the CBD. One of the clearest markers of this difference is attire. For both men and women managers the solid lines of the business suit can be read as an attempt to construct the professional body. This means delimiting bodily flows. But it would be incorrect to assume that the experience of men and women managers is the same simply because both wear professional attire. Grosz (1994a: 203) argues 'that women's corporeality is inscribed as a mode of seepage'. If business suits function to firm up the body then there is surely a significant point about gender to be made here.

> [W]omen, insofar as they are human, have the same degree of solidity, occupy the same genus, as men, yet insofar as they are women, they are represented and live themselves as seepage, liquidity. The metaphorics of uncontrollability, the ambivalence between desperate, fatal attraction and

> strong revulsion, the deep-seated fear of absorption, the association of femininity with contagion and disorder, the undecidability of the limits of the female body (particularly, but not only, with the onset of puberty and in the case of pregnancy), its powers of cynical seduction and allure are all common themes in literary and cultural representations of women.
>
> (Grosz 1994a: 203)

Grosz is not proposing that women are inherently or naturally more fluid than men but rather that women's bodies are constructed through discourse as 'modes of seepage'. Men's bodies also seep but their leakages are rarely represented. For example, after urination men's penises sometimes drip but this is something that is rarely spoken about.[2]

If, as Grosz suggests, women have been defined on the side of the body and liquidity and men on the side of the mind and solidity then it is possible that the business suit takes on an added significance and importance for women. The interviews indicated that a number of the women managers recognised this point. Veronica, a management consultant in Edinburgh in her mid-30s, explains:

> Women working as managers need to be taken seriously so how people look is important. It is part of the reason I dress the way I do. I started doing this work at 27 and anything that I do in terms of how I present myself which makes me look like I know what I'm talking about, to be taken seriously, helps enormously.

Veronica is acutely aware of the need to dress in a way that enhances her (solid) professional image.

Deborah Jones (1992) draws on feminist and semiotic approaches in order to analyse the coding of gender in clothing for managerial women. She concludes:

> Managerial women's clothing – from frilly blouses to padded shoulders – can be seen as the embodiment of a series of provisional resolutions of the conflict between the categories of 'woman' and 'manager'. Managerial women's clothing distinguishes them from the secretaries on the one hand, and from male managers on the other. . . . The performance of being a woman/manager, then, is grounded in paradox.
>
> (Jones 1992: 3)

McDowell (1997), after interviewing women and men merchant bankers in London, argues that women are 'marked' by what they wear. Men's attire, on the other hand, is the unmarked category. Women are judged harshly in relation to their clothing – it is too tight, too bright, too revealing, too masculine, too harsh. 'Whatever she wears, she draws attention to herself' (McDowell 1997: 145). In John Molloy's 1977 best seller, *The Woman's Dress for Success Book*, women are advised: 'NEVER wear anything sexy to the office' (capitals in original). However, if professional women completely emulate men (with no touches of the feminine such as a blouse instead of a stiff collared shirt, a skirt instead of trousers, small heels instead of flat shoes) they are often constructed as too 'hard', as 'butch', or as second-rate men. They become women cross-dressed as men and as such, are subject to ridicule. It is impossible for women to follow the ever changing and nuanced rules of dress conduct. To reiterate McDowell's (1997) comment, women are always marked by what they wear.

This gendered politics of dress was also evident amongst some of the participants who do not always wear suits. In cafés and health clubs, for example, suits are usually considered to be overly formal although managers in these industries do at times wear a suit. Liz, manager of an Edinburgh health club, explains:

> I wear whatever is smart – blazer and badge – but obviously to teach classes I'd get changed but then I'd go in the shower and put my suit back on. Some of the managers in the group will wear track suits like the staff but I prefer not to. If I'm going to be meeting people then I ought to dress as they expect me to dress and not be stuck in a track suit.

Liz's use of the phrase 'stuck in a track suit' is interesting. It echoes Kirby's (1992: 12–13) claim that Woman 'remains stuck in the primeval ooze of Nature's sticky immanence'. When stuck in the soft folds of her track suit Liz finds it difficult to be taken seriously as a manager. Being stuck – immersed in stickiness – alludes to solidity being privileged over viscosity and fluidity (Grosz 1994a: 194).

The women interviewees in this study were fastidious about following the (often unwritten) rules of professional attire. This is not surprising given that women are the marked category – constructed as 'naturally' prone to being leaky, messy, volatile bodies. Women managers seemed to find it more difficult than their male colleagues to disregard the normative practices surrounding professional attire. Out of a total of 26 interviewees (13 women and 13 men) only three men and no women claimed to disregard some, or any, of their workplace dress codes.

Guy, a 36-year-old manager of an Auckland Information Technology company, explains that most people in his position would wear a suit but he chooses not to.

> I'm not just selling what I do, I'm selling myself . . . I've got suits that I could wear but I don't like wearing suits so my sort of chosen style of presentation is (.) what would you call it? 'Smart casual', 'California

> casual' . . . There are catalogues that come out with 'California casual'. It's sort of open neck shirts, or if you're wearing a shirt, it's not a pressed business shirt, it's a linen shirt with a collar or tie, or it's a waistcoat that doesn't match a suit and it's a pair of linen pants or cords [corduroy fabric] . . . I'm not trying to present an overly up-market, or corporate, I want to be more on a comfortable level rather than on a too highly strung professional level with my clients. You know, it creates a better rapport, it's more for me. I think about it quite a lot actually because a lot of the people in my industry do wear suits and ties when they go out to see clients but I'm trying to present quite an up-beat, what do you say, like an 'arty' sort of image.

I quote Guy at length because he has clearly given a great deal of thought to the way in which he dresses. He is very aware of the rules that govern decency and professionalism in workplaces but chooses to push these boundaries a little. So too does Trevor, manager of an Auckland security company. The day I interviewed Trevor he was wearing black linen trousers and a cream linen shirt with a bold lime green tie. He describes his style as 'noisy corporate'. Trevor adds:

> I think my clothes reflect an image I want the company to have. You know I expect my guys [he also employs women] to present themselves well, and, you know, the best way to reflect that is to set the standard rather than drag it down [he laughs].

Mike, who describes himself as Irish/Scottish, aged early 30s, is manager of an Edinburgh high fashion retail business (part of a large European chain). The day I interviewed him he was dressed in boots, baggy linen trousers and a rollneck sweater. He was unshaven and had two earings in one ear. I remarked that other managers whom I had interviewed tended to wear suits. Mike explains:

> I didn't buy a suit this season. I don't have to, I mean, I get a discount and I can get a suit but this season I didn't want a suit 'cause it's a new store and it's, we carry a lot of bigger stock than we carried in our old store. And I wanted to represent some different looks as well and experiment. Over the winter I like to be sort of warmer as well. I don't like suits, you're too naked around the neck.

In an interesting reversal Mike associates the suit with feeling 'naked around the neck'. Thus far I have constructed suits as a form of body armour that firms up the body but for Mike the suit represents vulnerability, around his neck at least. Mike's job was to manage his staff by encouraging them to sell a range of 'looks' – not 'a business look' but a range of sexy, sensuous, casual looks. These 'looks' rely upon bodily surfaces and orifices appearing volatile and open (often read as inviting sexual and sensual pleasure) rather than stable and closed. Mike's earings, stubble, and soft polo-neck jersey adorned a managerial body that did not follow the normative practices of managerial dress. Nevertheless, Mike's purpose was the same as other managers I interviewed – to make as high a profit as possible for the company. Mike was clear about this. Mike's business, however, was to sell particular kinds of sexual and sensual looks and he was leading his staff and customers by example. None of the women I interviewed, even in fashion retail, dared to transgress managerial dress codes in this way.

Despite the fact that most managers wear suits, the 'proper' body can never be guaranteed. There are 'perilous divisions between the body's inside and its outside' (Grosz 1994a: 193). During some of the interviews the cracks in the presentation of the proper managerial body began to emerge. For example, when I interviewed Harry, manager of an Auckland outdoor recreation shop, his nose kept running. As mucus leaked from his nose into the stubble on his top lip he struggled to keep in control by dabbing it with a tissue. Jerome, an Auckland bank manager in his early 30s, in a conversation about some of the pros and cons of purchasing a suit through a retailer rather than through 'the bank wardrobe', 'admitted' that he had problems controlling his weight. Jerome explains: 'In between the time when you order it and the time of delivery, for the bank wardrobe, there are several months. I can change sizes in that time.' McDowell and Court (1994: 740) point out that weight seemed to be particularly significant for the youthful male merchant bankers in London. Changing suit sizes over the space of a few months can be read by others at the bank and/or by clients as evidence of a body that is not stable and in control.

Another body that was not completely in control was Robert's. Robert, a 52-year-old manager of a human resources consultancy in Auckland, talked about the grey hairs in his beard. He explains that some grey hair facilitates his work role ('If I was a 27-year-old I wouldn't even get in the door' says Robert), but too many just make you look 'old' (rather than 'mature').

Yvonne, manager of an Auckland health club, talked about being pregnant. When I asked her how she thought her clients might respond to her changing body shape she responded with some apprehension.

> Um (.) the pregnancy issue, we'll sort of find out about later, but you know, in the role you come up against some funny males, which is interesting, you know, you sort of think the world is changing but you wonder whether it is or not, you have a few bad ones, men who think they know it all.

In such conversations it became apparent that the managerial body (like all bodies) cannot be sealed safely in a business suit. The body that drips mucus, gains and loses weight rapidly, has greying hair or is pregnant attests to the impossibility of the 'proper' body. People do not always have consummate control over their bodies even in workplaces where such control is actively sought.

It also became obvious during the interviews that the two spheres of home and work are far from autonomous despite some of the managers' attempts to create a binary division between the two. Managers' bodies are attached both emotionally and corporeally to other bodies – those of partners, lovers, children and friends. Even though I did not specifically ask people about their home lives sometimes they began to talk about their 'loved ones'. During an interview, Astrid, manager of a home decor store in Edinburgh, began to question me about whether I had family in Edinburgh. Before long the interview transgressed its 'proper' boundaries as we conversed about toilet training our toddlers (whose bodies, as mothers, we were attempting to control).

Robert, manager of a human resources consultancy, explains that his daughter was completing a planning degree at Auckland University. Debbie, manager of an Auckland leisure complex discusses her difficult relationship with her partner: 'my partner finds it really hard . . . the only time that he sees me is like on a Sunday because that is the only day that I have off'. Peggy, an Edinburgh bank manager, explains that she is divorced and has a 'wee boy'. Public and private narratives merge. The 'cracks' begin to appear in the rational, respectable facade. It becomes evident that workplaces, and the bodies they interpellate, are fragile and volatile.

## Disciplining the corporate body

Managers are not only concerned with their own modes of dress and behaviours, however, but also with those of their employees – the collective or corporate body. Consequently, strict guidelines about staff attire are often put in place. In retail clothing stores, employees (and sometimes store managers) are often provided with, or have to purchase, items from the store to wear as their uniform. This allows customers to view a range of store clothing on living mannequins. Hilary, manager of an Auckland clothing retail store, explains:

> We get specific directions from Melbourne head office as to what we are allowed to wear for the season . . . And so they select the colour that we are to wear, which they believe is a reflection of the season. And they also profile, different stores can choose from different selections. So, for example, we're classed as a career store because we're in the inner city. So, yes, we have quite specific guidelines as to what we may choose from.

Staff need to have the 'right look'. Debbie, manager of an Auckland leisure complex employing 1,800 uniformed staff, also stressed this point. She comments:

> As far as what we wear, our uniforms, it comes down to marketing and the wardrobe department. Yes, it's a combination. It's a marketing issue though, because it's the image issue. It's who we are. It's a reflection of who we are as an organisation. So from a marketing perspective they have a huge input into, you know, what are our colours, where do other clothes go, what's the image we're wishing to portray. We have all that stuff to be concerned with . . . we have appearance policy standards for the whole company so, you know, the expectation is business attire. No matter where you work, we do not allow jeans for any staff, whether they're front of house, back of house. We don't allow tights [she is referring to fabric or thick 'leggings' rather than sheer nylon pantihose]. We have a very high standard of dress that we expect from all of our employees, managers as well. So yeah, I expect all my staff to be in suits. I would expect, you know, the ladies and the guys to wear a suit. And I wear a suit. It's just how it is.

Charles, in his late 40s and manager of an expensive and reputable menswear shop in Edinburgh, employs only three or four staff but he shares Debbie's sentiments about the need for staff to appear well groomed. Charles explains:

> In this shop I have no problem with it, the staff as regards to grooming . . . Stuart's been with the company for 20 odd years and he knows exactly what's required. If he needs to he polishes his shoes after coming to work in the morning. If it's been a muddy day outside, we would never go into a shop full of muddy shoes and things like that so we always keep facilities in the canteen there so to spruce ourselves up a bit. We just shine the shoes, wear [name of the company] clothing, neat and trim hair cut, and that's all we require. That's what the people expect to see when they walk in.

I asked Charles if he would consider employing someone with long hair or a tattoo. He replied:

> I wouldn't start interviewing anyone for a job with a tattoo. You know the type of people you're looking for and you wouldn't start in on that . . . A nice lad started with us about six months ago . . . when he first came out here he looked a bit scruffy but he agreed to follow the firm's traditions and he's done well and he's quite happy now.

Charles succeeded in disciplining the 'nice lad's' body. This theme of disciplining employees in relation to their appearance emerged in many of the managers' narratives. For example, Kathryn, manager of an Edinburgh branch of one of the United Kingdom's largest department stores, states:

> Well, you'd pick up on anybody who didn't look right quickly then. Make sure they didn't have scruffy jackets on. [At this point in the interview a trainee manager, Wendy, who is present comments humorously: 'And no nail polish'.] That's my, one of my things, someone in food section had management training and refused to take her nail polish off.

Debbie also had a great deal to say about the disciplining of her workers' bodies including the suitability of different bodies to different jobs.

> We have a number of large people because, I mean our wardrobe, for example, goes up to a size 26. So we have some quite large people who actually work here and by nature some of the jobs that we actually have require some fairly large people because of the weights that are carried. Well you don't need to be large to carry heavy weights but that has advantages. They will be big body shapes anyway, and therefore they tend to be big. And a lot of our staff are Polynesian staff and by nature they tend to, it's a generalisation, but they are larger than Europeans. And you know New Zealand Europeans are bigger than the Europeans from Europe. So, but yeah, we have a good mix and I think there are some areas that there will always be mangers who will say 'this is an image-conscious role' and therefore we tend to move towards people who are smaller body shapes.

This racialised discourse indicates that this company is more likely to appoint large-bodied Polynesians to poorly paid, invisible positions that require heavy lifting while smaller-bodied Pakeha are appointed to better-paid, more visible positions such as reception desk work. Debbie continued her narrative arguing that workers need to learn to present themselves properly.

> It's learning to present. I think there is a huge gap in the market for somebody who actually helps people learn how to dress . . . I know that there are places like 'Style' and [inaudible] and all those sorts of organisations but they tend to be driven by the wrong things. I think if they could get people in to actually show them how . . . and they could do it at a reasonable price then they'd get a huge market. We have somebody who comes in as part of our 'hiring celebration' that we call, that is actually a dress consultant

who actually comes in, talks to people and says 'Okay, well, this would suit you better', or, 'This is what you should do with your make-up to help you there', or, 'You really need to think about this'.

According to Debbie, employees' style of clothing and make-up is not the only issue. So too is body cleanliness.

DEBBIE: But we have issues, other issues, like people remembering to wash and that, and that just, that is something that is a *huge* problem in New Zealand.

ROBYN: Really?

DEBBIE: Yes, yes, a huge problem in New Zealand. I don't know what's gone wrong with society. It's more prevalent in New Zealand than anywhere else.

ROBYN: And what makes you think it is a New Zealand thing?

DEBBIE: I talk with other companies, international companies.

ROBYN: I see. And how do you deal with that?

DEBBIE: You have to tell them 'You smell' . . . You know, New Zealand Employment Services [now known as the Department of Work and Income] and I have had a number of conversations about this. They often say they have to tell people how to present themselves to go for a job and part of that will be, you know, make sure you have a shower [emphasis in original].

Thinking that Debbie's narrative might be part of a racialised discourse constructing some people as Others – 'dirty' and 'impure' – I asked if this 'problem' applied to both men and women, and to people from a range of ethnic groups. She replied:

Absolutely, absolutely! I'll find somebody who comes from a wealthy family as much as I'll find somebody who comes from a family which is on a lower income, that will have the same problem.

It appeared as though Debbie's comments applied equally to everyone, although after my question and her three line response she seemed reluctant to comment any further which left me with some doubts.

While most managers would not want their employees to have a strong body odour (although perfume on women is often considered acceptable) not all managers disciplined their employees bodies in exactly the same way. Mike, the manager who did not wear business attire, explained that he did not want his staff to have just one 'look', rather he wanted them to portray a range of 'looks'.

The main thing with the staff on the shop floor is to have a wide range of looks. You have like an India look, you have a fashion look, you have high fashion look, you have an androgynous look, you have the little girl look.

> It's got to be mixed . . . You don't want them to wear the same style hair. They get given a uniform. They get given clothes to wear, they pick them themselves but they have to be okay. You want a group of looks. They can't all be in jeans and T shirt. They can't all be in black boot cats and a little slinky top. You've gotta, like vary it. Someone's got to wear a dress, someone's gotta wear a skirt, someone's gotta wear trousers. In menswear someone's gotta wear a suit.

In Mike's narrative the suit is reduced to just one more 'look' that can be sold to potential customers. Despite the fact that Mike advocates a plurality of 'looks' for his staff there were still rules – regarding nails, shoes, body odour, hair, make-up jewellery and facial piercings – that staff are expected to adhere to. Mike explains:

> Appearance wise, if someone, if they bite their nails we would ask them not to bite their nails, because you do so much with your hands. You handle money, you handle garments, talk to the client. You quite often use your hands when you speak, you're telling them where things are. Horrible, scraggy bitten nails just aren't on. You can't have scabby shoes. Your shoes have got to match what you wear. Or, a bit of BO [body odour] or something, 'cause you're working with the public we can't allow that to happen. I mean I'm not a clean shaven person but you can't be like scruffily unshaven. We wouldn't like, your hair has got to be clean. Yeah, that kind of thing. We like girls, I mean we don't say you've got to wear full make up but we like, you know, they like, they would wear some mascara and some lipstick. Jewellery, we like them to wear jewellery. We like the girls to paint their nails. The guys can paint their nails if they want. That type of thing. Just, I mean nail polish, you know, it's like really, like big, big, big, big, big business in the last couple of years. Like instant fashion colours. It's constantly changing. I mean, it used to be you weren't allowed to have tattoos in this company. And you couldn't show them. But that's not really the case now. It's in my contract that you can't have facial piercings but lots of the staff have facial piercings now. It's an old, that's the reason I have this [points to several earrings in each ear].

All of the managers play an explicit role in disciplining the bodies of 'their' workers although the prescriptions for the ideal presentation of employees changes. Kathryn reprimanded one her staff members for wearing nail polish whereas Mike requested that his staff wear nail polish. Charles wanted his staff to look traditional and 'clean cut' whereas Mike wanted his staff to portray a range of contemporary looks. What remains constant throughout these narratives is the managers' unquestioned desire to shape not only their own 'look' but also the 'look' of their employees. Just as important as the clothes, however, is the flesh of the body – the comportment, gait, fat, hair, muscles, skin tone and colour, and facial features.

## Sculpting managerial flesh

Flesh that is firm rather than soft, bodies that are trim rather than flabby or obese are desired by most managers keen to portray a 'solid' company image. Bordo (1990: 90) points out that the contemporary ideal body is:

> absolutely tight, contained, 'bolted down,' firm (in other words, [a] body that is protected again eruption from within, whose internal processes are under control). Areas that are soft, loose, or 'wiggly' are unacceptable, even on extremely thin bodies. Cellulite management, like liposuction, has nothing to do with weight loss, and everything to do with the quest for firm bodily margins.

Many of the managers I interviewed expressed a desire not simply to be slim but to be firm, toned, fit and well exercised. This was especially the case in Auckland, even though the interviews took place in the middle of winter when early morning and evening temperatures can fall below freezing making it difficult to find the motivation to exercise.[3]

Perhaps the desire to be fit is not surprising given that being active and sporty, especially in the outdoors, plays an important role in the construction of national identity in New Zealand. Scott Crawford (1987: 161) argues that New Zealand is a country devoted to sport and the sport image has been an important foundation for the development of a 'national identity'. Sporting success in New Zealand is seen to forge national pride and is 'the most valued form of cultural achievement' (ibid.).

Colin, a commercial bank manager in his late 40s, explains: 'I find that I'm more efficient if I'm fit. As a consequence I keep myself on a fairly regular fitness programme and have done for the last 15 years or so'. Guy, an Information Technology manager in his mid-30s, states:

> Usually I'll go running, I'll do stretches but I'll usually go running and I'll run for about 20 minutes to half [an hour] and it varies . . . In the last two weeks I've started running a minimum of four times a week for half an hour. It's not a lot but it's enough to keep going.

Many of the managers work out in downtown gyms either before work, during the day or after work. Caroline, manager of an Auckland café, comments: 'I have a membership with a gym and a pool and I also cycle, So, I cycle in and cycle home.' Lex, manager of an Auckland downtown gym, combines his weekly exercise with work. He explains: 'Me and the director workout three times a week so we try and have our meetings while we're doing that [laughter]. So it's quite good. It works well.' Trevor, manager of an Auckland security company, explains: 'I do a reasonable amount of running from time to time I guess is the best way to describe it. And I go to the gym from time to time.' Trevor adds: 'Once a week I get a round of golf in the afternoon'. Auckland managers Debbie, Gillian and Lois also understand exercise to be an important part of their weekly activities. Debbie, explains that she travels to the gym at 5 in the morning, does 'a bit of circuit work' and then runs to her place of employment, arriving at about 7 a.m. Gillian states: 'I try two or three times a week to go the gym'. Lois explains that she plays netball in order to remain fit and 'de-stress' after busy days.

Four Auckland managers 'admitted' having not exercised much over the winter period. Alex, a bank manager, explains (with some degree of shame and embarrassment): 'Yeah, I was [exercising], over summer especially, I played a fair bit of sport, and over winter, I was playing rugby up until last season, so, over winter, no I haven't done much this winter at all'. Hilary, manager of a clothing retail store, similarly explains: 'Yes I do [exercise] and I have to admit I've been a bit lazy on these cold mornings [laughter]. There's not a lot of incentive when it's dark but as a rule my husband and I walk in the mornings. We probably go about three or four times a week but not the last month or so.' Harry, manager of an outdoor clothing and equipment store says that he used to do 'a fair bit of kayaking and tramping' and 'a little bit of climbing' but that he has not done a great deal of these activities since he started managing the shop. Yvonne, manager of a corporate health club, explains that she would feel as though she was 'doing an injustice to the members' if she was not very fit, trim and healthy. She adds to this, however:

> Having said that I now think you know that I'm pregnant [laughter] and so I haven't actually exercised for three months but I've got a good excuse for that at the moment! [laughter].

Managers in Edinburgh also exercised, although to a somewhat lesser degree than their Auckland counterparts. Bryan, manager of a travel agency in his early 20s, plays football on Sundays and goes to practices twice a week. Charles, in his late 40s and manager of a clothing retail outlet, attends line dancing several times a week. This keeps him 'pretty fit'. Astrid, manager of a furnishings and home décor store aged in her mid 30s, gets up at 5 a.m. most mornings to train for a triathalon.

> So I get up, we commute, so I leave the house at seven but I get up at about five and then I can go out running or cycle and swim if I have a lunch hour . . . And then I've got two horses that I look after as well, from when I was a teenager.

Lyn, personnel and training manager of a large hotel, had recently joined a gym.

> Yes. I have been [exercising] lately because I have to go in for a fitting for my wedding dress so I've joined the gym. Same [gym] as my fiancé. And we tend to go down say three nights a week. So I'm busy doing a bit of that. And I walk to work as well. I live past the castle so it's a 20 minute walk every morning.

Nigel, sales manager of a large department store, understands himself to be:

> quite sporty. I'm mad about sport. I enjoy golf, play a lot of tennis. I don't get as much time as I'd like to but I think most of the management I know, particularly the younger element, are quite involved. We tend to do a lot. People set up competitions, we do golf tournaments, we do five-a-side tournaments.

Some of the Edinburgh managers were less keen to exercise and found it difficult to muster enthusiasm on cold, dark mornings or evenings. Wendy, a trainee manager in her early 20s, says:

> I used to a lot more. I don't 'cause I find that starting at eight and finishing at six by the time you've travelled home and got your tea, the first thing I do is crash out on the sofa before I can go and do something else. Especially in winter.

Ruth explains: 'I don't exercise, no, not at all. Not at all. I keep saying I'm getting around to it but I never do'. Dick, manager of a bar, had bought himself a £500 multi-gym about six weeks ago but had only got around to using it about four times.

Working out in a gymnasium, jogging, cycling, swimming, playing racquet sport and so on are an integral part of daily life for many managers. Often these activities are heralded as good relievers of stress but they also function to craft a corporeal self that marks the individual and corporate body as 'in control' and 'successful'. Simon, manager of a downtown corporate gym in Auckland, explains:

> The guys want muscles. A lot of them don't want to build up in size but they just want the muscles to stand out a lot more so they can still wear a suit but they can look good in a singlet at the same time. So they're looking good *all* the time [emphasis in original].

Simon adds to this:

> It's a wellness of life that they're [professionals, managers etc] actually looking for at the moment. They want a better way of handling their stress, they want to feel good, they want to look good, and they want that energy so that they're buzzing every day . . . They want to look good because . . . it's like a big competition out there, what they are wearing, how they are dressing, they've got to be a hundred per cent to obtain that corporate image and it's hard to maintain.

Many of the managers with whom I spoke were very committed to keeping fit and healthy and to looking good. Dick, manager of a security firm, argued that clothing, body shape and grooming are extremely important because, after all, 'the perception of the fat slob in an expensive suit is still a fat slob in an expensive suit. It's not a nice thing to say but it's a reality.' Bordo (1990: 90) notes a contemporary disgust with lumps, bumps, bulges, fat, flab, cellulite, stomachs, guts – 'the soft, the loose; unsolid, excess flesh'. These, she argues, are unacceptable even on extremely thin bodies. In the next section I focus on one aspect of this trim, taut embodiment, that is, flexibility.

## Flexing managerial muscle

During the research the word 'flexibility' emerged so often that I was prompted to think further about the relationship between discourses of flexibility and the (in)flexible material bodies of managers. For example, in Auckland, Jerome explained that 'while being flexible is a good thing because you do learn to adapt to different things . . . [it] can also be dangerous because you do more'. Guy, who works in the information technology industry, used the words 'flexible' and 'flexibility' at least 15 times in his interview. Colin discussed flexibility in terms of people's ability to multi-task while Lois claimed that you needed to be flexible to work in the travel industry. In Edinburgh, Peggy talked about banks as flexible and working flexi-time. Nigel discussed the need to be flexible in terms of mobility, stating that managers had to be prepared to shift from city to city in order to progress their careers. Mike claimed that you have to be flexible about having a mix of men and women working on the shop floor.

The past decade has seen the emergence of a substantial literature on economic restructuring, paid work, organisational theory and gendered employment relations. Within this literature there is a related body of research on flexibility and flexible work practices. Arguments have been made that the efficiency of global capitalism depends on flexible labour. Much of this labour-related research focuses on flexible work practices, firms' demand for labour force flexibility and the consequences of this demand for workers (see Harvey 1989). A number of feminist analyses of labour force flexibility have also been carried out (Christopherson 1989, Fujita 1991, Reimer 1994).

The 1990s saw 'flexibility' became a buzzword in many countries, including New Zealand and Scotland. The *Collins English Dictionary* (1979: 555) defines flexible as '1. able to bend easily without breaking; pliable. 2. adaptable or variable'. Most of the economic and geographic literature on flexible work practices is in keeping with this second meaning – adaptable or variable – variability of working hours, independent sub-contracting, temporary work and part-time work. There has been less discussion of managers and workers needing literally to flex, to have bodies that are pliable, limber and supple.

Kantor (1989: 361 cited in Martin 1994: 210) makes the comment that American (and one could add here New Zealand and Scottish) corporate bodies in the 1990s aim to be 'focused, fast, friendly and flexible'. The individual bodies that make up these corporate bodies are also being encouraged be 'focused, fast, friendly and flexible'. I am not simply talking about 'mental flexibility' here (as though the mind can be divorced from the body) but also 'bodily flexibility'.

OCOM Office Communication Limited (a company that sells faxes, computers, phone systems and copiers) advertises: 'We bend over backwards for you'. I have seen their advertisement on a roadside billboard and in several business magazines and newspapers in New Zealand (for example in *Waikato Business News*, 1 July 1996). Accompanying the advertising slogan is a photograph of a professional man in his mid-20s, dressed in a dark-coloured business suit with a patterned tie. The man is completely bent over backwards in a U-shape as though he is about to do a backwards somersault. Such corporeal acrobatics are beyond most of us yet this man also manages to hold a cell phone in one hand, support himself completely on his other hand, and smile at the camera (see Plate 5.4). In short, like the company he is representing, he is 'focused, fast, friendly and flexible'. What is represented in this advertisement is not so much an ability to solve problems in a manner that is flexible, or someone who is necessarily skilled at range of tasks, rather, it is flexibility at the level of the material body itself.

There are numerous examples of similar advertisements. The Edinburgh Carphone Warehouse's Winter catalogue 1997/98 contains an advertisement for a Mitsubishi Electric Mobile Phone, which depicts a man in his mid-20s, wearing a suit, business shirt, tie and roller blades. He is holding a briefcase in one hand and a mobile phone in the other. He is smiling and suspended mid-air against a backdrop of corporate towers. He appears to be fit, athletic, trim and flexible.

Martin (1994: 212) claims that in the United States (and I would argue also in New Zealand and Scotland) large corporations are requiring workers and management to participate in new kinds of experiential training methods. Martin was invited to attend a day-long session run by a training company for the employees of a multinational corporation. 'Twenty-two thousand Rockford employees were in

*Plate 5.4* Billboard advertisement for OCOM Office Communications Limited 'We bend over backwards for you'.
*Source:* Photograph by Robyn Longhurst, 1996.

the process of going through three days of workshops as well as high and low ropes courses at a rural site on a large bay on the East Coast'. Martin (1994: 212) explains:

> Protected by sophisticated mountain-climbing ropes and harnesses, teams of men and women workers and managers of all ages and physiques (as well as Karen-Sue and I) climbed forty-foot towers and leaped off into space on a zip line, climbed forty-foot high walls and rappelled down again, climbed a twenty-five-foot high telephone pole, which wobbled, stood up on a twelve-inch platform at the top, which swivelled, turned around 180 degrees, and again leaped off into space. (This last is privately called the 'pamper pole' by the experiential learning staff because people so often defecated in their pants while trying to stand up on it.)

Clearly such acts require bodily flexibility. Martin (1994: 213; my emphasis) reflects: 'I was literally moving from one position of instability to another and experienced the necessity for *great flexibility*'.

Interviews indicated that small and large companies in New Zealand and Scotland are using experiential training. Just outside of Auckland is the Clevedon Confidence and Adventure Centre; in Manukau City (located on Auckland's southern boundary) is the Experiential Training Company (ETC) – to name just two. In Edinburgh, Peggy, a bank manager, explained that she had taken part in an experiential training course run by Bedlam:

> It was like an outward bound 'Crystal Maze' [a television programme screened in Scotland]. We had to build bridges and put up an abseiling thing and stuff like that . . . it was an absolute nightmare and we were shouting at each other . . . the hardest thing we had to do was we were all blindfolded and we had to follow the rope and listen to each other's instructions and stuff like that.

Peggy was not especially committed to outdoor pursuits or to exercise in general and so the activities that she had been forced to engage in proved to be 'an absolute nightmare'. Lyn, manager of an Edinburgh hotel, is fit, slim and exercises at least several times a week. Every January she engages in team building exercises. Lyn explains:

> We've done abseiling and we've done rock climbing and we've done a lot of games, you know the 'plane drops' – the plane has dropped three packages and you've got to find them in the radius – and all that sort of thing. So, it was really great.

Many of the managers with whom I spoke, in both Auckland and Edinburgh, are expected to dress 'appropriately' and also to have a flexible (trim, taut and terrific) body – a body that can abseil and rock climb. Such a body appears to have become an unwritten requirement for managers in the late 1990s. Businesses want managers (and workers) who are able to 'risk the unknown and tolerate fear, willing to explore unknown territories . . . In a word, *flexibility*' (Martin 1994: 214; emphasis in original). Representations of flexibility are being 'lived' at the level of the material body.

Another point that emerged in the interviews in relation to flexibility and the body is that it is only possible to flex something to a specific point of tension, beyond which it breaks. When talking with managers in Auckland and Edinburgh it became evident that 'stress' (mental and physical) is a concern. Some bodies are being flexed or stressed to a point where breaking becomes a possibility. For example, Jerome, an Auckland bank manager, talks about the increase in his stress levels over the last few years.

> The workload has been building and building and building over and over the past three to five years I guess, so all of that comes into it as well, and

> the job itself is very stressful of course. You're dealing with money matters and you're often growling at people because they're not operating things the way that they should be operating. So, there is a lot of conflict.

Auckland managers involved in this research worked, on average, 52.3 hours per week. Edinburgh managers reported working fewer hours, on average, 46.5 hours per week. Long hours are one possible reason why some managers suffer stress. Debbie talks about hitting a wall.

> There's no doubt about it, it is stressful. You can't work that number of hours on a continual basis [Debbie works, on average, 70 hours per week] and be effective all the time. And there are periods when I know I am not as effective as I could be because I've just got up to a wall . . . you hit a wall and you say 'I've really gotta have space and I've gotta have time out and I've gotta have a week off or whatever'.

When I asked Guy if stress was an issue in his job in the Information Technology industry he responds: 'Definitely, definitely'. I also asked Sandra whether stress was an issue in her job as manager of a café. She replies: 'It can be very stressful if you're not in control'. Alex explains that he is not the only one in his office to suffer from stress.

> Over the last probably 18 months or so there's been several people in this office who, that I know of, probably other offices as well, who've been affected with stress-related illnesses. Just basically the health seems to deteriorate a wee bit and you go to the doctors or whatever and they suggest taking a week off work not because there's anything, I guess, particularly viral or anything wrong with you as such. It's just to get back down again.

Jerome's use of the phrase 'get back down again' is interesting. His was one of the first interviews that I conducted. At the beginning of the interview process I was not sure what he meant by 'back down'. After conducting more interviews it became apparent that many of the Auckland managers desired a state of high energy levels, to feel energised, to be 'up', to be flexed, to be 'buzzing'. Lex, manager of a downtown Auckland health club says that their clients (mainly 'professional' people, including managers) 'want a better way of handling their stress, they want to feel good, they want to look good, and they want that energy so that they're buzzing every day'. In this sense, Jerome's phrase 'it's just to get back down again' makes sense.

Yvonne, manager of a very expensive downtown Auckland health club, comments that most of their clients come to the club in an attempt to 'decrease blood pressure, increase aerobic capacity which therefore increases energy . . . the reason they come is usually to help decrease stress levels and to combat blood pressure'. Yvonne explains that her club offers 'stress management services' but they are 'looking at changing that name because . . . people actually tend to be put off by the word "stress"'. Lex, manager of another downtown Auckland health club, also remarked that some of his clients 'who come after work are so stressed out they need something before they actually go home'. In fact, says Lex, 'Most people that come here are people who are pretty stressed'.

## Gender, flexibility, fluids

Although both men and women managers voiced desires to feel and look 'good' at work, this discourse has a gendered dimension. Women's bodies are often considered to be 'naturally' more pliable, supple, bendable, limber, yielding, adaptable and compliant – in short, more flexible, than men's. Aerobics, gymnastics, ballet and yoga classes at most gymnasiums and recreation centres in New Zealand and Scotland are dominated by women who stretch, twist and bend in an attempt to exercise, recreate, prepare for pregnancy/childbirth and, in general, increase their bodily flexibility. Men are stereotypically seen to be hard, strong and tough, but not necessarily flexible. I began to wonder if this new discourse might assist women's entry in managerial positions. If having a flexible body is considered an attribute for managers then women may be perceived as having some 'natural' advantage. This is not to suggest that other attributes such as education, self-confidence, intelligence, competence on the job, hard work, the 'right' clothes and hairstyle will no longer count for anything, rather having a flexible body needs to be added to this list of performance criteria. This reading of women being more flexible than men and therefore enjoying an advantage in the managerial labour market, however, is perhaps too simplistic.

In both New Zealand and Scotland, men are commonly regarded as the most competitive and proficient at sporting and physical activities. They are thought to run faster, play tougher and, I think, more recently, to possess greater bodily flexibility. The advertisements in New Zealand and Scotland, by and large, do not portray women bending over backwards or roller-blading. They portray urban men who are fit, trim, well toned, well groomed, fast, friendly, and flexible – not 'kiwi blokes', rural farmers or Scottish crofters.

Constructions of masculinity are changing in New Zealand (see Law, Campbell and Dolan 1999) and in the United Kingdom (see Jackson, Stevenson and Brooks 1999 and McDowell 1997). New Zealand has traditionally been seen as a place where men could 'reassert their masculine identity through physical work' (Phillips 1987: 17). This work was usually carried out in rural rather than urban places. Lawrence Berg (1994: 251) argues:

> It was commonly felt that the city was effeminate because it was a place of voluptuousness and fashion, of luxury and ease, where men lived a soft life . . . Cities were also inhabited by office workers, not 'productive' farmers and labourers . . . Cities and towns were for women then; the frontier was the true place of the 'kiwi bloke'.

However, as Rose (1991: 159) suggests 'masculinist discourse is complex and differentiated' – it continually recreates itself in ways that make it resilient to women gaining power easily. Masculinist discourse in relation to managerial culture in New Zealand is recreating itself. 'Rural kiwi blokehood' appears to be giving way to the 'plastic Queen Street fellas' (these phrases were used in the *Heartland* documentary series on Tokoroa screened by Television New Zealand in 1995). The bodies of the 'Queen Street fellas' are trim, flexible and well 'suited' to high-performance work and play.

Some of the conceptual differences attached to men's and women's bodies may be just as important in understanding flexibility. Throughout this book I have argued that men's bodies are often coded as firm while women's bodies are coded as soft and as 'modes of seepage' (Grosz 1994a: 203). Women's bodies are understood in relation to the fluids or viscous material that emanate from within them. Irigaray (1985) argues that fluids tend to be implicitly associated with femininity, maternity, pregnancy, menstruation. Fluids are subordinated to that which is firm, concrete and solid.

In order for a body to be flexible it must be firm (but not completely rigid). The body that is thought constantly to threaten to 'leak', 'seep' and break its boundaries cannot flex – it has in essence already collapsed in on itself. It makes sense to talk about a telephone cord as flexible but it does not make sense to talk about honey, water or blood as flexible. Therefore, given the conceptual codings of bodies, it is difficult for women to compete with men in relation to flexibility. What might appear to be a level playing field for men and women managers turns out to be anything but level. What might appear to be biological bedrock turns out to be anything but a common, universal humanity. Fleshy material bodies cannot be plucked out of discourse.

In this section I have dealt with just one variable in relation to flexible bodies, that is, gender. It is, however, important to consider some of the implications of this discourse of flexible bodies for 'other' bodies, for example, bodies that are ill, impaired or disabled (see Dyck 1995 on women with multiple sclerosis and Chouinard and Grant 1996 on women with disabilities). Access to, and mobility in, the CBDs of Auckland and Edinburgh is very difficult, and in many cases, impossible, for people who are ill, impaired or disabled. The vertical corporate towers, uneven roads and pavements in Auckland, cobbles in Edinburgh, crowds of people, heavy and revolving doors, are uninviting, if not downright prohibitive, to many.

Bodies that are not flexible cannot 'make it' in such disabling environments. It appears that to be a manager working in the CBD it is necessary to have a particular kind of body that can engage in physical activity. This discourse of flexibility clearly has implications for people who are ill, impaired or disabled. My observations indicate that there are few such people in managerial positions in the CBDs of Auckland and Edinburgh.

It is likely that there are also implications of discourses on flexibility for bodies of 'different' colours and ethnicities. All the participants in this study had 'white' skin. There is a powerful imperialist legacy that constructs 'white' bodies as respectable and 'proper'.

> To be respectable means to belong to a 'civilised' people, whose manners and morals are more 'advanced' than those of 'savage' or backward people. In this schema people of colour are naturally embodied, amoral, expressive, undisciplined, unclean, lacking in self control.
>
> (Young 1990a: 138)

It is highly likely that recent discourses of flexibility are not only privileging able-bodied men but also that whiteness is retaining its dominant position (see Rose 1993a: 33 on whiteness denying its own colour, see also McClintock 1995). Although this topic lies beyond the scope of this chapter I suggest it is a useful direction for future research.

Likewise there are many questions about sexuality and workplace performances that remain unanswered in relation to fit, firm and flexible bodies. McDowell and Court (1994) examine the (hetero)sexual imagery used to portray young, white merchant bankers in London. Similar lines of enquiry about the hegemony of heterosexuality could be pursued in relation to managers working in CBDs.

Young (1990a: 124) argues that 'the objectification and overt domination of despised bodies' has receded and there is now a 'discursive commitment to equality'. However, Young continues that racism, sexism, homophobia, ageism and ableism, have not disappeared. Instead they 'have gone underground, dwelling in everyday habits and cultural meanings of which people are for the most part unaware' (ibid.). While some groups are unconsciously defined as ugly and abject, others are unconsciously defined as attractive and desirable. Managers are appointed, partly, on their status as attractive and desirable. Young (1990a: 139) argues:

> It is paradigmatically in the office, or at business meetings, that persons in contemporary society follow the rules of decorum typical of bourgeois respectability, and in these settings people evaluate one another according to these rules.

Examining, and even changing, policy will not necessarily change the way in which this unconscious privileging of specific bodies takes place. It is unwritten rules that govern the most diminutive aspects of our everyday, banal, existence. Bodies are 'thought' to require discipline and vigilance and only those who achieve this succeed as managers in the CBDs of Auckland and Edinburgh. The managers' reward is not only social acceptability and an ample salary package but also the power to discipline and manage the bodies and spaces of others.

# 6

# SOME THOUGHTS ON CLOSE(T) SPACES

In this concluding chapter, I draw together some thoughts on pregnant bodies in public places, men's bodies in toilets/bathrooms, and managers' bodies in workplaces in CBDs in order to illustrate that bodies and spaces are neither clearly separable nor stable. I attempt to destabilise notions of self/other and subject/object in relation to these spaces. I slip between talking about the body as a space (for example, the interuterine space of the pregnant body) and the intimate spaces that the body inhabits (for example, domestic toilets/bathrooms). The spaces of the body and its environs become close, intimate, merged and indeterminable as they make each other in fluid and complex ways. The interuterine spaces of pregnant bodies, defecating men and managers whose bodies attempt, but inevitably fail, to be respectable – conjure up images of close(t) spaces. They are close spaces in that they are familiar, near and intimate. They are also close*t* spaces in that they are often socially constructed as too familiar, near, intimate and threatening to be disclosed publicly.

As close*t* spaces they function as sites of oppression and resistance (see Brown forthcoming on 'closet spaces' and Sedgwick 1990 on epistemologies of the closet). Homosexual practices are often closeted, so too are a range of other bodily practices. Women are sometimes closeted about being pregnant – 'coming out' as pregnant can be both exciting and traumatic. In this book I have exposed the water closet in academic discourse. I have also discussed many managers' attempts to remain closeted behind the doors of respectability.

I do not mean to imply that a binary division ought to be drawn between 'in (closet) spaces' and 'out spaces', close spaces and 'far away' spaces, the body and the nation, the local and the global, the micro and the macro scale, views from above and below. Soja (1996: 314) argues that '[s]uch binarizations . . . are never enough'.

I focus on close(t) spaces not out of a sense of voyeurism but because they are as 'political' as any other ('far away' or 'out') spaces. The instability of boundaries, whether they be the bodily boundaries of individuals or the collective boundaries of nation-states, causes anxiety and a threat to order. To ignore close(t) spaces is to ignore that which is coded as intimate, 'queer', feminine, banal and Other. Such a strategic absence allows masculinism to retain its hegemony in the discipline. Close(t) spaces need an opportunity to come out in geography. There are many censoring and discriminatory practices that operate to keep particular sights/sites in the closet. Bell (1995) argues that articles are pulled from library collections. Secretaries sometimes refuse to type or copy certain material. There are whispers and silences from colleagues and negative press from the media. Editors have been known to refuse to publish material in geographical journals because it is 'inappropriate' (read: they are repelled by and fearful of the material). An editor of a well-known geographical journal once told me that the pregnant body is an 'inappropriate' subject for geographers to consider.

I am not alone in attempting the further understanding of close(t) spaces – spaces that are not clearly self or Other, subject or object. Homi Bhabha makes an argument for Third Space. Bhabha (1994: 39) claims that 'by exploring this Third Space, we may elude the politics of polarity and emerge as the others of our selves'. In an interview on the Third Space, Bhabha (1990: 211) explains that for him the Third Space is hybridity. Hybridity, he explains, is 'a process of identifying with and through another object, an object of otherness, at which point the agency of identification – the subject – is itself always ambivalent, because of the intervention of that otherness' (ibid.). In addition to Third Space, Bhabha (1994: 38; emphasis in original) uses a number of other spatial metaphors to articulate his notion of hybridity – these include 'alien territory', the 'split-space of enunciation' and '*in-between* space'. Edward Soja, inspired by Henri Lefebvre's *The Production of Space*, has devoted an entire book to *Thirdspace* (1996) in which he encourages readers to think differently about space and spatiality.

Others have found the notion of what Plato in the *Timaeus* calls the *chora* to be useful in retheorising space and spatiality. The *chora* (which Kristeva refers to in order to explain her notion of the semiotic) is 'receptacle, unnameable, improbable, hybrid, anterior to naming, to the one, to the father and consequently maternally connoted' (Kristeva 1980: 133). The *chora* is the site of the undifferentiated bodily space the mother and child share. 'A site for the production of the matrix/womb and matter, the *chora* is the unnameable, unspeakable corporeality of the inextricably tangled mother/child dyad which makes the semiotic possible' (Wright 1992: 195).

Currently, there are a number of academics including geographers Gibson-Graham (1997) and Sharpe (1999), sociologists Lechte (1993) and feminist theorist Grosz (1994b) who are using the notion of *chora* to further understand issues of space, place, architecture and sexual politics. Gibson-Graham (1996) discusses the possibilities of thinking a postmodern pregnant space and notes that the inherent femininity of *chora* lies in its immanent productiveness. Gibson-Graham (1996: 90) refers not only to the *chora* but also to: 'the third space of political choice' depicted by Soja and Hooper (1993: 198–199) (drawing on Foucault's [1986] notion of heterotopia) which is a place of enunciation of a 'new cultural politics of difference'. Gibson-Graham (1996: 90) also refers to Rose's (1993a: 137–160) discussion of a 'politics of paradoxical space' and de Lauretis's (1986: 25) comments on 'elsewhereness' and 'space-offs'.

For the purposes of this chapter I was tempted to use one or a range of these ideas on Third Space, the *chora*, and paradoxical space because they convey a sense of ambiguity, hybridity and ambivalence. In the final instance, though, I desired a notion that would speak more directly to a feminist politics of intimacy, fluidity, viscosity, mess and dirt. Moss and Dyck (1999: 389) argue for what they call 'corporeal space' 'where the discursive and the material are synchronous'.

> Corporeal space consists of context, discursive inscriptions, material – economic and matter-based – inscriptions, the biological, and the physiological . . . These spaces are fluid, congealing from time to time around the body, only to be destabilized with new boundaries forming when any part of the context, the discourse, or the materiality shifts.
>
> (Moss and Dyck 1999: 389)

I think the idea of 'corporeal space' is potentially rich for considering body space relationships because it conveys something of the fluidity and messiness of bodies and spaces. 'Corporeal space' is in keeping with what I have chosen to call close(t) spaces.

Despite the fact that '"The body" is becoming a preoccupation in the geographical literature' (Callard 1998: 387) it is still difficult to speak of close(t) spaces, liminal zones, abject bodily sights/sites in the discipline. These spaces threaten to spill, soil and mess up, clean, hard, masculinist geography. Codes of respectability place limits on what we can say in geography. We may be able to discuss discursive constructions of embodiment but we still cannot talk easily about the weighty materiality of flesh, or the fluids that cross bodily boundaries in daily life.

The close(t) spaces of the pregnant woman/uterus, of toilets/bathrooms and of supposedly respectable bodies and workplaces are both real and imaginary. They are spaces of tears/blood/sweat and spaces of discourse and representation. The pregnant woman is both self and Other, mother and fetus, one and two, subject and object. The defecating man is also both subject and object. His excrement is both of him and distant from him. Likewise the manager who attempts to remain respectable at all times at work inevitably gives way to belching, burping or farting. S/he is both a respectable self and a loathsome Other. It is worth pursuing each of these ideas in turn.

## Interuterine spaces

During pregnancy the zone or space around the body changes. The zone around the pregnant stomach becomes considerably thinner and may even disappear altogether in some instances. Interpersonal relations are situated within the multiple discourses that surround pregnancy and come into play to create a new spatiality for pregnant women and for those who interact with them.

This new spatiality helps to make sense of the public touching of pregnant women's stomachs. It can also be understood in terms of the pregnant woman herself who at times is no longer sure where her body begins and ends in relationship to the geographical space that she occupies (see Young 1990a: 164). This can lead to a sense of uneasiness, surprise and disjuncture between the image and the materiality of the body for pregnant women. The pregnant woman sometimes finds her body in places where she does not expect. For example, she may try and squeeze through a gap only to find that her stomach protrudes further than she thought. Consequently she is unable to 'pass'. The pregnant subject's anatomical, material body can grow rapidly and it often takes some time before her body image catches up. As the pregnancy proceeds the borders of the body image change – they 'are not fixed by nature or confined to the anatomical "container," the skin' (Grosz 1994a: 79).

This point becomes particularly evident when examining various understandings of the placenta. 'Facts' relating to the structure of the uterus, especially the placenta, have changed radically since the 1960s. Sharon Marcus (1993: 135), in a fascinating paper on *Rosemary's Baby* (a novel, which was later made into a movie), discusses how, 'In the early 1960s, obstetricians viewed the placenta as a barrier that guaranteed the autonomy of the fetus within the womb and protected it from any maternal influence'.

Although there is an implied link between mother and fetus, they were largely understood to be almost entirely separate entities. The 'wall' of the uterus functioned to protect the fetus. Gould (1958: 11 cited in Marcus 1993: 135) writes:

> we are blessed that our children are so well protected from all that we swallow or inject into ourselves. Every child has a silent nurse, constantly on duty, protecting him [sic] from almost every kind of poison, and much better prepared than his mother to see to it that he receives precisely the right amount of food and oxygen. This nurse screens out almost everything in the mother's bloodstream that could harm the child, before it reaches him, and even manufactures the extra chemicals that he needs. That nurse is the placenta . . .

In the first half of the 1960s it was thought that the womb insulated the fetus from the mother and 'is not too different from the space capsule designed to sustain and protect astronauts in outer space' (Hughes 1964: 22 cited in Marcus 1993: 135). This understanding of the placenta allowed for a sharp distinction or separation to be drawn between mother and fetus. The pregnant woman's body was seen as 'outer space' with the fetus being located in a sealed or walled interior – 'space capsule' – known as the womb.

This model of the placenta-as-barrier or wall meant that not only was the fetus thought to be independent from the mother, but, also that the mother was understood to be independent from the fetus. This obviated some (but not necessarily all) of the present day pressures on pregnant women to monitor or circumscribe their activities in order to safeguard their baby. Pregnant women were not expected to give up smoking, employment, sport, to take plenty of rest, to curtail activities outside the home – at least not on the grounds that these activities might threaten the fetus in some way. Pregnant women may have been pressured to give up these activities on other grounds, for example, it was not seemly or lady-like for pregnant women to engage fully in public life, but not on account of the well-being of their fetus.

It was not until a 'discovery' early in 1965 'that the placenta facilitated rather than blocked communication between mother and the fetus' (Marcus 1993: 135) that a new discourse leading to an increase in the surveillance of pregnant women (both self-surveillance as well as surveillance by others) began to emerge. A 1967 July column in *McCall's* noted that 'the old idea that the womb is the safest human habitat has been sharply disproven in recent years' and cautioned that 'infants' were now being 'attacked in the womb' (Kerr 1967: 48 cited in Marcus 1993: 136). Marcus (1993: 135–136) suggests that this is a:

> more overtly paranoid version of the placenta . . . paranoid in its representation of the pregnant woman as the persecutor of the fetus . . . [and] paranoid in a psychoanalytic sense, since it depicted the pregnant woman's body as unable to maintain the distinction between self and other, mother and fetus, exterior and interior.

Marcus (1993: 137) claims that writers for women's magazines and medical professionals interpreted 'the new view of the placenta to mean that mothers could communicate infectious or harmful substances to fetuses. As a result, those authors cautioned mothers to practice constant self-surveillance.' This self-surveillance involved getting plenty of rest, giving up employment, reducing activities outside the home and relying 'on housework to stay physically fit' (Marcus 1993: 136–137). This idea of pregnant women's behaviour directly affecting the fetus still reverberates today. Marcus (1993: 137) explains:

> The new view of the placenta led to arguments that because women's

> reproductive organs enabled a dangerous communication between the internal fetus and the surrounding 'maternal environment,' the mother could only protect the fetus from her own dangerousness by assiduously exercising paranoid self-surveillance and transforming herself into an agoraphobic invalid.

In order to suggest a different and new understanding of the placenta and interuterine environment (that is, an understanding that could prove mutually beneficial for both mother and fetus) it is worth considering the work of Hélène Rouch – a biology teacher at the Lycée Colbert in Paris. In an interview in 1983 (cited in Irigaray 1993: 38) Luce Irigaray asks Rouche: 'can you explain the mediating role the placenta plays during interuterine life?' Rouch begins her answer by explaining the placenta. She says: the placenta is

> a tissue formed by the embryo, which, while being closely imbricated with the uterine mucosa remains separate from it . . . although the placenta is a formation of the embryo, it behaves like an organ that is practically independent of it. It plays a mediating role on two levels. On the one hand, it is the mediating space between mother and fetus, which means that there is never a fusion of maternal and embryonic tissues. On the other hand, it constitutes a system regulating exchanges between the two organisms, not merely quantitatively regulating the exchanges (nutritious substances from mother to fetus, waste matter in the other direction), but also modifying the maternal metabolism: transforming, storing, and redistributing maternal substances for both her own and the fetus' benefit. It thus establishes a relationship between mother and fetus, enabling the latter to grow without exhausting the mother in the process, and yet not simply being a means for obtaining nutritious substances.
>
> (Rouch cited in Irigaray 1993: 39)

Rouch is not so much describing a biological reality as an imagined reality that is culturally determined. Rouch understands the placenta as continuously negotiating differences between the self (the mother) and other (the embryo). She claims that the relative autonomy of the placenta cannot be reduced either to a mechanism of fusion, or, conversely, to one of aggression. In a psychoanalytic and material sense, the uterus/fetus is the other of the pregnant woman's self.

Recognising this interuterine space as a close(t) space may offer a way of reconceptualising the relationship between mother and fetus but also of reconceptualising space more generally. All of us have occupied interuterine space (it is perhaps the closest of all spaces) and yet it is seldom discussed in geographical discourse (it has long been closeted). It has been closeted because the maternal, fluid, indeterminate geography of the uterus is likely to mess up a masculinist knowledge based on claims to truth, objectivity and rationality.

At a broader level it is useful to recognise that all spaces (not only interuterine spaces) are a 'bleeding' of materiality, fluidity, the imaginary, the socially constructed and the psychoanalytic. Bodies as/and spaces are always already soiled – they are never self-contained but can only exist in a complex relational nexus with other bodies/spaces.

## Excremental spaces

Like the pregnant body, the defecating or excreting body that both constitutes and is constituted by domestic toilets/bathrooms is often understood to be a potentially dangerous body. In the usually private space of toilets/bathrooms people not only excrete but also check their appearance in the mirror, check their weight, pass wind, wash and comb their hair, shower, bath, clean teeth, shave, squeeze pimples, masturbate and rub on creams. Toilets/bathrooms are also a site where subjects might attend to the bodily needs of others (for example assisting children and/or the elderly with toileting). Such activities are not discussed widely by geographers.

There is something about toilets/bathrooms and the (excremental) bodies that they house that constructs them as an 'unspeakable space' – a close(t) space – in geographical discourse. Toilets/bathrooms tend to be considered too material, too squeamish, too uncomfortable, too unacceptable (or just plain too banal) to discuss. Toilets/bathrooms have not been considered as a material or discursive space in which bodies are imbued and inscribed by cultural practices. They have not been on geographers' research agenda.

Kayrn Kee, a librarian at the University of Waikato, searched a variety of data bases including five First Search databases, two general databases, a New Zealand database and three databases currently on trial at the University of Waikato. Key words such as 'bathroom', 'toilet', 'lavatory' and 'water closet' were used but yielded very few 'hits'. Using various search engines on the World Wide Web also yielded very little information on domestic toilets/bathrooms. There are some sites on the design of bathrooms, planning ideas, demolition and remodelling, ceramic tiles, vinyl flooring, cabinets, marble countertops, showers and toilets but their aim is to promote bathroom products.

A great deal more geographical research has been carried out on *public* (rather than domestic) toilets/bathrooms. Examples of this work on public bathrooms (sometimes known as restrooms) include Stephen Hodge's (1995) research on public toilets as a possible 'beat' for gay men. Hodge (1995: 45) argues: 'The ambiguous nature of beats emanates from their overt function as lavatories and places of recreation while covertly being sites of sexual activity'. Peter Keogh (1992) discusses urinals as a site of cottaging for gay men. Sally Munt (1998:

776–780) offers an insightful account of being a 'butch' in the 'Ladies toilet' and the boundary disruptions this causes. Robin Law, Annabel Cooper, Jane Malthus and Pamela Wood (1999) present a historiography (from mid-nineteenth to the mid-twentieth century) of public toilets in Dunedin, New Zealand. Julia Edwards and Linda McKie (1997: 135) examine the provision of public toilets for women beginning with the question 'why is it that women invariably have to queue for the toilet in public places whereas men do not?'

Interestingly, in these examples the focus tends to rest on gay men or women, rather than on heterosexual men. Few geographers have focused explicitly on heterosexual men. Some notable exceptions are Peter Jackson (1991), who discusses the use of black men in advertising (but these men's bodies are Othered by their 'blackness' – see Mohanram (1999) on 'black bodies'), Gregory Woods (1995), who examines 'island narratives' that function to construct 'natural' (read: heterosexual) masculinity, Linda McDowell (1995), who discusses male (and female) merchant bankers and their engagement in heterosexual gender performances, Doreen Massey (1998), who also focuses on relationships between masculinity, home and workplaces examining a binary between reason and non-reason in scientific research sectors of high-technology industry in Cambridge, England and Glendon Smith and Hilary Winchester (1998) who draw on interviews with 25 members of men's groups in Newcastle, Australia, to examine men's attitudes to, and cultural practices of, work and home. By and large, however, there is no geographical discussion of heterosexual men in the private spaces of toilets/bathrooms.

The lack of attention paid to domestic toilets/bathrooms is all the more interesting in the light of a substantive geographical literature on the home. Over the last two decades feminist geographers have problematised the dichotomy between public space (exterior and open) and private space (interior and closed) and the ways in which public space has been valued over private space. As a result of this critique the home has increasingly become a legitimate topic for consideration. Some useful examples of feminist work on the home include Dolores Hayden (1976, 1978) on envisaging utopias including kitchenless houses, Leslie Kanes Weisman (1992) and Matrix (1984) on domestic landscapes as 'man-made', Louise Johnson (1992, 1994b) on suburban housing, and Lynda Johnston and Gill Valentine (1995) on lesbians' experiences of 'heterosexual family homes' and 'lesbian homes' in the United Kingdom and New Zealand.

Within this literature there is some mention of specific rooms, for example, the kitchen is mentioned as a site of women's labour. Laundaries, bedrooms and 'family' rooms are also mentioned but toilets/bathrooms rarely feature in this feminist geographical literature on the home (although Johnston and Valentine 1995: 106 note the experience of a lesbian who was nervous about her parents using the bathroom when they visited her flat because this would have forced her 'to come out'). It would appear that the toilet/bathroom is even more closeted and removed from the scrutiny of others than the more public areas of homes such as lounges and kitchens.[1]

I do not want to suggest, however, that toilets/bathrooms are completely absent from the geographical imagination. Saarinen (1976) discusses the design of toilets/bathrooms in homes drawing on Alexander Kira's study *The Bathroom* (1966):

> With a team of researchers at Cornell University, he [Kira] analyzed the varieties of behaviour that take place in bathrooms. They measured heights, reaches, breadths, ranges of movements, and other physical characteristics of people and tabulated various uses of the bathroom and each fixture in order to design the facilities better.
>
> (Saarinen 1976: 21)

Saarinen (1976: 22) makes the point that 'often psychological and cultural factors must be considered as well as the physical dimensions and physiological functions'. He argues that when considering the design of a bath it is important to consider not just its efficiency in relation to its 'cleansing function and ease of getting in and out' but also how comfortable it is, that is, its function as a place to relax. Saarinen's study indicates that geography's boundaries are porous, they leak and seep. Elsewhere (Longhurst 1995b, 1997) I have argued that the body is never entirely absent in geographical discourse. Nor are domestic toilets/bathrooms entirely absent. Excluding the body (and toilets/bathrooms as an abject site/sight) 'is not the whole story of masculinism' (Rose 1993a: 6). In order to establish the mind (and its associations with rationality, and public spaces such as boardrooms) there must be a contrast with the body (and its associations with irrationality, and private spaces such as toilets/bathrooms).

Even though toilets/bathrooms have not been completely excluded they have most certainly been largely ignored – treated as banal and unimportant. It is likely that they are considered commonplace by many geographers – unworthy of attention – but they are also sites/sights of threat. They are one of geography's abject and illegitimate sites that have been deemed (perhaps unconsciously) inappropriate and improper by the hegemons in the discipline. Munt (1998: 78) describes toilets as 'liminal zones'. It is difficult to speak of liminal zones. These zones, and their articulation in language, may cause us to feel uncertain, uncomfortable, confused and/or maybe repulsed. Liminal zones are often unspeakable.

One of the reasons for a dis-ease and dis-comfort over what are considered to be abject sights/sites in geography is the privileging of the mind over the body in geographical work. The body tends to be Othered in geographical discourse (Longhurst 1997: 486–501). Questions, therefore, about the omission of particular sites/sights in geography (for example, toilets/bathrooms) slip into questions about rights (women's rights, rights to be 'queer' etc). This Othering of toilets/bathrooms serves to marginalise certain individuals and groups (such as women, the disabled and so on who are thought to be 'tied to their bodies' and, therefore, incapable of reason). It is a specific notion of knowing as disembodied that marginalises Others in the production of geographical knowledge.

The mind/body dualism plays a vital role in determining what counts as legitimate knowledge in geography. Geographies of toilets/bathrooms run the risk of being ghettoised, feminised geographies robbed of their legitimacy. So long as the mind is privileged over the body, hegemons in geography will continue to edit out that which they consider to be 'dirty', preferring instead the clean, the clinical, the quantitative, the heroic, the solid, the straight and the scientific. What constitutes appropriate issues and legitimate topics to teach and research in geography comes to be defined in terms of reason, rationality and transcendent visions, as though these can be separated out from passion, irrationality and embodied sensation. Unlike boardrooms, toilets/bathrooms are not considered to be the 'real' or 'serious' stuff of geography.

Toilets/bathrooms are (abject) zones or sites where bodily boundaries are broken. The insides of bodies make their way to the outside (for example, urination, excretion, vomiting, squeezing pimples) and what is outside the body may make its way to the inside (for example, the naked body may feel vulnerable to penetration). The emotions that can accompany these acts remain sealed within the privacy of toilets/bathrooms. A danger lurks here. They are a site/sight that bears testimony to 'the fraudulence or impossibility of the 'clean' and 'proper' body' (Grosz 1994a: 194). They are the sites that house the 'daily attributes of existence' (ibid.).

Focusing on heterosexual 'white' men's experiences of toilets/bathrooms in Chapter 4 is a way of rewriting male corporeality. In the spaces of toilets/bathrooms men cannot pretend to make any easy separation between subject and object, self and Other, desire and repulsion, solid and fluid, mind and body. Cartesian ontology is 'messed up'. The fluidity and permeability of heterosexual, 'white' men's bodies is exposed. Focusing on the toilet/bathroom forces heterosexual men to come out of the (water) closet.

## Managing spaces

Unlike toilets/bathrooms, 'professional' workplaces, especially in CBDs, are constructed as spaces in which bodies must not transgress their boundaries. Liminal zones where the insides and outsides of bodies sometimes become indeterminable – noses, vaginas, penises, eyes, sores – must be carefully monitored and kept under control at all times in workplaces. It can require enormous vigilance to construct the proper, professional and respectable body – to present a 'public face' – at work. This is one of the functions of the business suit.

The firm and straight lines of the business suit give the appearance of a body that is impervious to outside penetration. They also give the appearance of a body that is impervious to the dangers of matter that is inside the body making its way to the outside. The suit closets the body in respectability. However, although the suit helps to create an illusion of a hard, or at least a firm and respectable body that is autonomous and in control, bodily boundaries can never continually remain intact. While business attire may reduce potential embarrassment caused by any kind of leakage it can never completely secure a body. Given that women function as the 'marked category' and that their bodies are socially constructed as 'modes of seepage' business attire takes on an added significance and importance for women.

While some workplaces are constructed as rational and cerebral, (for example, banks, computer consulting companies or insurance companies) others are constructed as pleasurable and fun (for example, cafés, casinos and health clubs).[2] Regardless, there is an assumed respectability in all these environments. Young (1990a: 137) notes that: 'The environment in which respectable people dwell must also be clean, purified: no dirt, no dust, no garbage'. Managers, staff and even customers who do not conform to the unwritten rules that govern respectable behaviour in these spaces are treated with suspicion and caution. There are continual attempts (by managers but also by cleaners and security guards) to keep professional workplaces clean, tidy and 'nice' regardless of whether they be a clothing store, bank, restaurant or travel agency. Clean, tidy, 'nice' sites/sights help to make clean, tidy, 'nice' bodies and visa versa. Together they ensure profits for a wealthy middle class.

The boundaries separating the 'nice' from the shabby (bodies and places) are insecure and continually contested. The corporeal boundaries of '"poor bodies" on the streets, under the stars' (Peace 1999: 444) leak and seep (their sores, cracked skin, stained clothes). These bodies, like queer bodies, disabled bodies and black bodies, engender feelings of abjection for respectable bodies. Respectable bodies, however, need an Other upon which to found their (insecure) identity (see Rose 1993a). Steve Pile, drawing on the work of Stallybrass and White (1986) makes the point that there is an internal relationship between power, desire and disgust. 'Others are kept at a distance . . . boundaries can only be achieved through constant vigilance: the powerful are constantly looking with desire and disgust, in fascination, at things which are considered outside their selves' (Pile 1996: 176). Sibley (1995) also makes this point in *Geographies of Exclusion*. Drawing on object relations theory Sibley examines boundaries at a range of spatial scales including at the level of the body and psyche.

## Conclusion

In all three case studies – pregnant women in public spaces, heterosexual 'white' men in domestic toilets/bathrooms, and managers in CBDs – ideas about sexed

›dies, body boundaries, body fluids, abjection and (im)pure spaces have proven useful. In concluding I want to make a series of brief points in relation to the ideas which have emerged in this book.

The first point is that it is not enough to examine only broad and wide-sweeping maps of power and meaning. The micro-level politics that imbue bodies and spaces also need to be held up to scrutiny. The body is as 'political' as the nation-state. As a generalisation, geographers have been effective at looking at the broader picture but this has sometimes been at the expense of finer detail – the close(t) geographies. In order to understand the relationships between people and places it is necessary to address a range of geographical scales. To date, the body has received less attention than more macro-level analyses.

The second point I want to make concerns the production and politics of academic knowledge. Sibley (1995: 185) advocates that 'geographers go out into the world [although he recognises that some are already there], not on an imperialist and colonialist mission, but in order to experience the lifeworlds of other people'. In this book I have attempted to do just that. I have talked with and listened to people while at the same time trying to understand myself (including my academic practice). During the years of writing I have been prompted to think about the production and politics of academic knowledges. I have no doubt that some readers will attempt to dismiss, suppress and/or neglect the ideas contained with in this book as illegitimate. David Sibley (1995: 121) argues that '[t]he issue of what, within the academy, counts as legitimate knowledge is a complex one'. Sibley notes that it matters *who* produces the knowledge (for example 'black' and female academics). The sites of knowledge production also matter.

I work as an academic geographer 'down under' at a small, largely unknown university in New Zealand (the 'bottom' of the world).[3] I often find that theoretical frameworks, usually produced in the Northern hemisphere, do not travel easily and unproblematically into specific Antipodean locales (see Berg and Kearns 1998; Mohanram 1999, Peace, Longhurst and Johnston 1997). This book then, may be read as a kind of hybrid, impure text that subverts Euro-Anglo theory with Antipodean narratives – a coming together of 'higher' and 'lower' knowledges. To put it rather badly – it has been written from the 'bottom', about the 'bottom'. My hope is that it will prompt readers to question further some of the grounds on which geographical knowledges rests and to consider the directions and constitutions of new geographies.

The third and final point that I want to make is that the construction of the body as Other appears to have changed throughout the 1990s as geographers began to examine more explicitly the politics of embodiment and spatiality. However, often the body being examined is a kind of 'disembodied body'. Irigaray (1993: 174 quoted in Davidson and Smith 1999) claims that women are concerned with a corporeal geography whereas men establish new linguistic territories. I think that in the discipline of geography over the last decade a new linguistic territory has been created. It is the territory of the body but ironically it is a fleshless territory, a territory constituted of little more than a chain of polite signifiers. It is a body that does not have specific genitalia or that breaks its boundaries. Such a body is a masculinist illusion.

Focusing on a body that has no specified materiality will not further feminist agendas, rather hegemony will retain its dominance. Denying the weighty materiality of flesh and fluid will help enable masculinism (the unmarked norm) to retain its hegemonic position. In this regard, the epistemology and ontology of geographical knowledge is unlikely to change dramatically.

In examining pregnant bodies in public spaces, men's bodies in toilets/bathrooms, and managers' bodies in CBDs I have illustrated that what all these bodies share is their fluidity, volatility, and abject materiality. Although some bodies are commonly represented as more abject that others (such as pregnant bodies), in fact all bodies (including the bodies of heterosexual, 'white' men) are unstable. In addition to this point I hope to have shown that this fluid, volatile, abject corporeality cannot be plucked from the spaces it constitutes and is constituted by.

The close(t) geographies of the body challenge some of the dominant constructions of knowledge in geography. Specificity seeps into generality, a politics of fluidity seeps into a politics of solidity, and a lived messy materiality seeps into cerebral knowledge. Perhaps thinking, writing and talking about bodily fluids, abjection, orifices, and the surfaces/depths of specific bodies can offer a way of prompting different understandings of power, knowledge and social relationships between people and places.

# APPENDIX: THE FIELDWORK

## General comments

The fieldwork was carried out separately for each of the three case studies – pregnant bodies in public places, men's bodies and bathrooms, and managing managerial bodies. In all instances participants were given pseudonyms. For the most part the research process that was undertaken is outlined in the relevant chapters. The focus of this Appendix is largely the analysis of data.

In all three case studies I was guided by Matthew Miles and Michael Huberman (1994: 10) in the three components of data analysis. The first component is data reduction. Miles and Huberman (ibid.) point out:

> data reduction occurs continuously throughout the life of any qualitatively orientated project. Even before the data are actually collected . . . anticipatory data reduction is occurring as the researcher decides (often without full awareness) which conceptual framework, which cases, which research questions, and which data collection approaches to choose.

The process of reducing data, and even *anticipating* reducing data, cannot be separated out from analysis. The second component of analysis is data display. Miles and Huberman (1994: 11) define display as 'an organised, compressed assembly of information that permits conclusion drawing and action'. As with data reduction, the use of displays is a type of analysis. Miles and Huberman (ibid.) claim that the third stream of analysis – conclusion drawing and verification – starts from when the researcher first collects data. From the outset the qualitative analyst decides

> what things mean – is noting regularities, patterns, explanations, possible configurations, causal flows, and propositions. The competent researcher holds these conclusions lightly, maintaining openness and scepticism, but the conclusions are still there, inchoate and vague at first, then increasingly explicit and grounded
>
> (Miles and Huberman 1994: 11)

Drawing conclusions, however, is only part of the story. These conclusions need to be verified.

I did not use a computer software package for analysing qualitative data. It is possible that a program such as NUD•IST (a package designed to aid researchers in handing *N*onnumerical, *U*nstructured, *D*ata by supporting processes of *I*ndexing, *S*earching and *T*heorising) could have assisted me. However, I decided against the use of such a computer package for several reasons (see Peace 1997 for a geographical perspective on using computer-assisted qualitative data analysis software). First, the quantity of data I collected, although substantial, was not overwhelming. Although a little bulky, it was possible to take home my folder of transcripts each night and to read them through. Also, I was able to recall most of the conversations without too many problems. Second, I collected most the data, and transcribed the bulk of the data, myself. Therefore, I was very familiar with the material. Third, computer-aided textual analysis systems such as NUD•IST are no substitute for a thorough understanding of social theory. That is, such programs do not do the analysis for the researcher, but merely aid analysis. They allow the researcher to conduct complex textual searches in order to build a hierarchical index system of interrelated ideas about the text. This is very useful but it is a tool which in itself is not capable of analysis.

## Pregnant bodies in public places

At the outset of the project on pregnant bodies I conducted preliminary interviews thereby beginning the process of 'anticipatory data reduction'. It became clear during these interviews that the two women who were pregnant for the first time, as opposed to the three women who were pregnant with successive children, seemed the most acutely aware of their changed corporeality as they began to confront what it meant to be a 'mother'. I decided to delimit the study by talking only to those women who were pregnant for the *first time*. I also decided to talk only with women who *lived in Hamilton*. The three women who had been pregnant several times, inevitably (no matter how carefully I directed the conversation) ended up drawing on their experiences of their other pregnancies which sometimes occurred in places other than Hamilton. I was also prompted to reduce the data I would collect by interviewing only women who were *visibly pregnant*. In all the interviews, others' reactions to the women's pregnant bodies, once people 'could tell', were discussed.

In general I did not dramatically reduce the data after collection. Rather, I anticipated quite specifically which data I would collect. For example, I did not reduce the interviews and focus groups at the point of transcribing because discussions had remained largely 'on topic'. I transcribed all of them in full rather than selected sections.

At the end of each dialogue transcription I wrote a précis of the interviews and focus groups and attempted to identify patterns that were emerging. Writing précis is a way of reducing and analysing data. I also wrote three vignettes (see Miles and Huberman 1994: 81–83 on researcher-produced vignettes) which perform a similar function. Writing a précis of each interview and each focus group, plus writing three vignettes, provided me with a way of reducing, summarising and analysing events and narratives.

I displayed my data in a number of forms. First, I had approximately 400 pages of transcripts in the form of narrative accounts about pregnant women's experiences. I transcribed all the interviews, with the exception of seven, which were transcribed by Lyndell Johns. I read through the transcripts at least three times and during those readings I noted both similarities and differences in terms of the topics that were discussed. For example, in nearly every interview with a pregnant woman some discussion emerged about her staying home more now that she was pregnant. I searched for networks, regularities and patterns in the data. Second, the four pregnant women who were involved in the study throughout their pregnancies kept a journal. They each gave this journal to me after they had given birth. Third, during the final stages of the project I myself became pregnant (with my second child) and kept a journal. I also draw on this experience in the chapter. Fourth, I analysed other forms of qualitative data such as brochures, newspaper clippings, magazine articles and medical forms that pregnant women are required to fill in. I also compiled a ring-binder folder of illustrations and photographs of pregnant women (taken and published by other people): all these were collected during the period 1992 to 1995. I read these newspaper and magazine columns, and looked at the visual images many times.

In order to verify the themes that I thought were emerging from the data, I conducted a computer search for key words and phrases. Since all the transcriptions were stored in one computer folder and I was able to search for words and phrases that were used repetitively by respondents. I used the word processing package Microsoft Word 5 to do this. The aim of searching for key words and phrases was to check out my 'hunches', to get a sense of the sturdiness, or otherwise, of my still inchoate conclusions. The themes that I thought were emerging were: 1) that women tend to go out less when they are pregnant; 2) that pregnant bodies are frequently represented (by themselves and others) as ugly, fragile, seeping and dangerous; and 3) that pregnant women are frequently represented (by themselves and others) as overly emotional and forgetful (I do not deal with this third theme in this book but see Longhurst 1996b). Therefore, I searched for words and phrases such as 'home', 'stay home', 'go out', 'fat', 'ugly', 'attractive', 'forgetful', 'irrational' and 'crying'. Note that I did not just search for words and phrases that would confirm my tentative conclusions, I also looked for words and phrases that might disprove the patterns that I thought were emerging.

## Men's bodies and bathrooms

In my preliminary conversations with people it became clear that men and women experience their toilets/bathrooms and talk about their toilets/bathrooms in different ways. Many women were able to talk freely for lengthy periods about particular body products, cleaning practices, the pleasure of relaxing in showers and baths. Men talked less freely. This information that I drew from the preliminary conversations was useful in helping me to refine and delimit my project. I decided to examine the experiences of only Pakeha, heterosexual men because they appeared to be the group who were least able to discuss their bodies in relation to toilets/bathrooms. The preliminary conversations also prompted me to reduce the data that I would collect in the future by talking only with men who lived in Hamilton. Conversations about toilets/bathrooms in rural spaces (for example 'out-houses'), in flats in London, in homes in Apia or in public spaces alerted me to the fact that comparing and contrasting experiences of a range of toilets/bathrooms in different places would be an entire research project on its own.

I had approximately 100 pages of transcripts in the form of narrative accounts of men's experiences of toilets/bathrooms. I transcribed all the interviews reading each one approximately three times. During these readings I noted both similarities and differences in the topics that were discussed. For example, in nearly every focus group there was a discussion about shaving. I searched for networks, regularities and patterns in the data. I developed 'hunches' about patterns that I thought were emerging, for example, that men are very comfortable talking about shaving but not about other bodily acts such as masturbation or passing runny stools.

Apart from the transcripts, which were the main source of data, I also analysed other forms of qualitative data. I compiled a ring-binder folder of interior decorating brochures and magazines, and newspaper and video clippings on toilets/bathrooms. It became apparent that many photographs of bathrooms depict 'clean' lines and colours such as blue – a colour long associated with cleanliness and purity. Toilets/bathrooms are rarely photographed with inhabitants. Grouping and displaying all the photos of toilets/bathrooms in one folder was useful in helping me to think further about bathrooms as clean/dirty spaces.

In order to verify the themes that I thought were emerging from the data, I conducted a computer search for key words and phrases. The themes I thought were emerging were: 1) men who share bathrooms with other men fear contamination through body fluids and products; 2) men aim to reduce the time they spend in domestic bathrooms; and 3) men can freely discuss shaving (bristles are solid) but find it more difficult to discuss masturbation and self examination. I also searched for words and phrases such as (bathroom and body) 'products', 'revolting', 'quick', 'time', 'minutes', 'shave', 'masturbation', 'pleasure' and 'mirror'. I also searched for words such as 'relax', 'enjoy', 'cleanse', 'moisturising cream'. I re-read many of the transcripts looking for conversations or references to men's pleasures in bathroom, their feelings of comfort in bathrooms. The fact that I could track very little information on such topics helped to validate what I thought were some of the emergent themes.

## Managing managerial bodies

I analysed over 200 pages of interview transcripts (most of which I transcribed in full). I also collected brochures, newspaper clippings, magazine articles and advertisements that depicted managerial culture. These were placed in a ring-binder folder. I noted the similarities and oppositions between the images of managers – their clothing, pose, facial expressions, and the 'bits' of their bodies that were included in each image. Displaying these data in ring-binder folders aided the process of comparing and contrasting information because I was able to continually regroup data depending on my thinking at the time. For example, it became apparent that many photographs of managers were aimed at making them look 'active' (in motion) and flexible. Grouping and displaying all the photos of these managers and reading the transcripts several times helped me to think further about the constructions of (normative and idealised) bodies.

As with the other case studies, in order to verify the themes that I thought were emerging from the data, I conducted a computer search for key words and phrases. The themes I thought were emerging were: 1) managers' behaviours at work are carefully disciplined to comply with codes of respectability; 2) the business suit performs an important function in firming (insecure and potentially dangerous) bodily boundaries; 3) managers discipline their own bodies but also the bodies of their employees; and 4) having a firm, fit and flexible body is increasingly becoming an unwritten requirement for many managers in CBDs. I searched for words and phrases such as 'professional', 'fit', 'firm', 'flexible', 'trim', 'groomed', 'clean' and 'exercise'. I also searched for words such as 'fat', 'tired', 'dirty', 'lack' and 'unfit'. In short, I looked for information that I thought might prove and/or disprove some of the themes that appeared to be emerging.

# NOTES

## 1 BODILY OPENINGS

1 From this point on in the text I use the term 'geography' but I am actually referring specifically to 'human geography'. Peter Haggett in Johnston, Gregory and Smith (1986: 205) suggests that: 'The internal logic of geographical study has tended to split the field into two parts: "physical geography" and a geography of human creations termed "human geography"'. For definitions of the terms 'human geography' and 'geography' (since a definition of 'human geography' cannot be divorced from the problems of defining the term 'geography' itself) see Johnston, Gregory and Smith (1986: 175–178 and 205–207).

2 Throughout this book I use the term 'bodies' to refer to people's bodies – human bodies. In general, 'human' geography has not yet extended its boundaries to include the bodies of 'non-human' animals but see Anderson (1995) on 'Culture and nature at the Adelaide Zoo'. Anderson develops a cultural critique of the zoo as an institution that inscribes various human strategies for domesticating, mythologising and aestheticising the animal universe. In relation to *human* bodies there is, of course, no one body – *the* body is a masculinist illusion. There are only bodies in the plural. Much feminist discussion focuses on the complex processes through which female and male bodies are differentiated. Bodies are *sexed*. Moira Gatens (1991a: 82) explains that 'the metaphor of a human body is a coherent one, but of course it's not. At least I have never encountered an image of a *human* body. Images of human bodies are images of either men's bodies or women's bodies' (emphasis in original).

3 The notion of discourse is complex, mainly because there are many conflicting and overlapping interpretations from a range of disciplinary and theoretical standpoints (see Fairclough 1992). Basically, however, two general understandings of discourse can be identified. First, in linguistics, discourse tends to refer to examples of either written or spoken language. Second, in social theory, discourse tends to be understood in broader terms. Although social theorists recognise the centrality of language to understanding discourse, they use the term to refer to ways of structuring knowledge and social practice (Fairclough 1992). Barnes and Duncan (1992: 8) characterise discourses as 'frameworks that embrace particular combinations of narratives, concepts, ideologies and signifying practices, each relevant to a particular realm of social action'. I use the term 'discourse' in its broad sense drawing on the work of Michel Foucault (for example, 1970, 1979, 1980). 'Foucault analysed the ways in which apparently objective and natural structures in society, which privilege some and punish others for non-conformity, are in fact "discourses of power"' (Bullock, Stallybrass and Trombley 1988: 232). A key aspect to Foucault's notion of discursive power is that such power is not just *repressive* (in that it prevents action) but that it is also *productive* (in that it creates new actions). Another important idea associated with discursive power is that it operates everywhere rather than just through specific agents. This includes at the level of the body.

4 I use the term Other (and Otherness) to refer to Jacques Lacan's (1981) concept of the Other. The Other

> refers essentially to the SYMBOLIC order of language and speech, [it] does not have a single meaning; it allows for more than one reading, and must be rigorously distinguished from the concept of other – with a small 'o' – which designates the relation to the specular other, the other who resembles the self, an imaginary relation which originates in what LACAN in 1936 called 'The mirror stage', and which describes the relation of the child to his image'
>
> (Marie-Claire Boons-Grafé in Wright 1992: 296)

In opposition to this specular other, Lacan proposed the Symbolic Other, which he paired with the Subject. Lacan (1981: 309 cited in Wright 1992: 298) defined the Other in 1955 as 'the place where is [sic] constituted the I who speaks with the one who hears'. Lacan linked this notion of Other to 'lack'. (For a fuller definition of Other see Elizabeth Wright's 1992 excellent dictionary *Feminism and Psychoanalysis*; see also de Beauvour 1953 on the role played by women as Other in psychic and social life.)

5 For a definition of the term 'masculinist' see Gillian Rose (1993a: 4), who claims that 'geography is masculinist'. She adopts the term from Michèle Le Doeuff (1991: 42 cited in Rose 1993a: 4) who describes 'masculinist' as 'work which, while claiming to be exhaustive, forgets about women's existence and concerns itself only with the position of men'.

6 Rose (1993a: 17) notes that: '[f]or white feminists, one of the most oppressive aspects of everyday spaces is the division between public and private space. One of the earliest discussions of the public and the private was an essay by Kate Millett published in 1969 and her arguments show how many feminists have connected the public/private distinction with patriarchal power.' This understanding of the distinction between public and private space as a debilitating one for women is also clearly evident in the work of many feminist geographers (see, for example, Matrix 1984; Weisman 1992). Most of these geographers have stressed the extent to which women's movements in public space are constrained by the ideological claim that women's space is the private domestic arena.

Rose (1993a) has also noted that the separation between the two spheres – public and private – often employed by feminists and feminist geographers is commonly a white one. Feminists of colour, such as P. Hill Collins (1990: 58 cited in Rose 1993a: 126) have 'detailed some of the reasons why the public and the private may not be appropriate terms for interpreting the social geography of Afro-American communities. The private was not always equivalent to the domestic home, for example, '"private" could refer to black community spaces beyond the reach of white people, both men and women. The private could thus be a resource for women – not a burden' (Rose 1993a: 126) (see also hooks 1990; Pratt 1984).

7 Aotearoa is the Maori term for what is commonly known as New Zealand. For more than a decade, especially since 1987 when the Maori Language Act was passed making Maori an official language, the term 'Aotearoa' has been used increasingly by various individuals and groups. For example, all government ministries and departments now have Maori names that are used, in conjunction with their English names, on all documents. Throughout this text I mainly use the term New Zealand but I want to alert readers to the fact that the naming of place is a contestatory process (see Berg and Kearns 1996).

## 2 'CORPOREOGRAPHIES'

1 The term 'corporeography' is used by Vicki Kirby (for example, in *Telling Flesh: The Substance of the Corporeal* 1997: 9). The earliest reference to the term that I can find is a chapter entitled 'Corporeographies' (Kirby 1989: 118). Kirby argues: 'the body is that "pre-post-erous space," the site of a corporeography that conjoins the dynamic political economy of signification – its written surface and writing surface'.

2 Elizabeth Grosz (1989: xvi) explains that when a continuous spectrum is divided into discrete self-contained elements which exist in opposition to each other, a dualism is created. 'Dualism is the belief that there are two mutually exclusive types of "thing" . . . that compose the universe in general and subjectivity in particular' (Grosz 1994a: vii), for example, white/black, good/bad, rich/poor, light/darkness, mental/physical, mind/body.

3 See, for example, Bordo (1989, 1993), Braidotti (1989, 1991), Butler (1990, 1993), Davis (1997), Foster (1996), Foucault (1980, 1985, 1986), Gallop (1988), Gatens (1996), Grosz (1995), Grosz and Probyn (1995), Haraway (1990), Irigaray (1985), Jaggar and Bordo (1989), Kirby (1991, 1997), Probyn (1993), Scott and Morgan (1993), Shilling (1993), Weiss (1999) and Young (1990a, 1990b).

4 As Vicki Kirby (1997: 138) notes, however, although this mind/body dualism is acknowledged and criticised in much postmodern writing this does not mean that the division itself can simply be overridden or rejected 'as if a quick condemnation of Descartes will afford the critic a protective talisman against the evils of dualisms'. It is not possible to simply step outside of Cartesian dualist thinking (although see Bhabha 1994: 36–39 on the 'Third Space' and Kristeva 1980 on 'the *chora*'). Nor is it particularly useful to reduce either the mind to the body or the body to the mind since it 'leave[s] their interaction unexplained, explained away, impossible' (Grosz 1994a: 7).

5 See Johnston, Gregory and Smith (1986: 358–360) for a definition of positivism in geography and Johnston (1986: 11–55) for more in-depth information.

6 See Johnston, Gregory and Smith (1986: 207–210) for a definition of humanistic geography and Johnston (1986: 55–96) for a more extensive review of humanistic approaches to geography.

7 Topophilia refers to people's affective ties with their material environment and particularly with specific places (Johnston, Gregory and Smith 1986: 493). According to Tuan (1974) it 'couples sentiment with place'. Johnston, Gregory and Smith (1986: 493) claim that:

> The term [topophilia] seems to have been first used by G. Bachelard in *La poétique de l'espace*[. The] concept is closely akin to J. K. Wright's Geosophy in that, through its focus on both Environmental Perception and on cultural values or attitudes, it necessarily concentrates on the study of places as either carriers of emotionally charged feelings or as perceived symbols. A slightly modified use of topophilia, closer in spirit to Bachelard's, occurs in the work of Edward Relph (1976) where it is taken to imply 'an encounter with place that is intensely personal and profoundly significant'.

8 It is ironic that male bodies tend to be missing from feminist scholarship. In a sense this has further accentuated the dualistic conceptual alignment of women with the body, femininity and irrationality, and men with the mind, masculinity and rationality. Some notable exceptions are Morgan (1993) and Connell (1995), but much work remains to be done in the area of men's bodies.

9 Psychoanalytic theory tends to concentrate on the development of individual identity from infancy onwards. Freud and Lacan, the two major psychoanalytic theorists of the twentieth century, have each identified a number of stages through which individuals pass in the

process of acquiring a sense of self separate from objects in the environment. Predominant in their theories is the problem of the male child's separation from the mother. Identity, for both these theorists, is indistinguishable from gender identity. Feminist psychoanalytic theorists, such as Nancy Chodorow (1978), have used this as a point of departure from Freudian and Lacanian approaches and argued that these theories too easily assume the primacy of male developmental processes. Despite this criticism, however, a number of feminists have considered psychoanalytic theory 'worthy' of building on for feminist purposes and a body of work known as 'psychoanalytic feminist theory' has emerged (see Wright 1992 for some useful insights as to whose work tends to be read as 'feminist psychoanalysis').

10 The term 'Pakeha' refers to Aotearoa/New Zealand born people of European descent. Although the term Pakeha has been (and at times still is) highly contested in New Zealand (see Spoonley 1993) it is now used as a standard term of classification of ethnicity in the New Zealand Census. The term Maori is commonly used to refer to the *tangata whenua* (literally 'people of the land') or indigenous peoples in New Zealand. I use this term here but wish to problematise such use. As Spoonley (1993: xiii) points out, 'the word "Maori" is really a convenience for Pakeha to lump together divergent groups'.

## 3 PREGNANT BODIES IN PUBLIC PLACES

1 I have chosen to call these maps symbolic maps yet many geographers may recognise them as a hybrid form of mental or cognitive maps used by humanist (especially behaviouralist) geographers in the 1960s and 1970s (see Downs and Stea 1977; Gould and White 1974; Lynch 1960; Tuan 1975 on mental mapping). It is worth noting that Map A has been drawn *retrospectively*. There is a possibility that pregnant women's *memories* of their life prior to pregnancy may well be different from their perceptions *at the time* of pregnancy. This does not mean that a map drawn prior to pregnancy would have been any more 'useful', 'accurate', 'truthful' or 'real', rather, it simply may have been different from the retrospective map.

2 Some pregnant women experience nausea and/or vomiting especially in the first few months of pregnancy and usually in the mornings (although it can occur at any time during the day, or at night). This is referred to as 'morning sickness'. Sometimes morning sickness is known as nausea and vomiting in pregnant (NVP). See Birks's (1993) book entitled *Coping With 'Morning Sickness'*.

3 The choice of transcription system is closely related to the type of analysis being attempted. As Ochs (1979) points out, transcription is already a form of analysis. It simply does not make sense to speak of the accuracy or completeness of a transcript without some framework for deciding which features of conversation are relevant or valued (see Cook 1990). At the same time, the form of transcription cannot be separated from practical constraints: to transcribe the many hours of recorded material I collected using one of the most comprehensive of systems would have been beyond my resources, as well as making the text difficult to follow for those readers unfamiliar with the system.

My concern was primarily with the content of discourse. I was less interested in the moment-by-moment conversational coherence of the interviews and group discussions. For this reason, I adopted a cut-down version of the set of conventions that have been developed by Gail Jefferson (1985) for conversation analysis. The transcription codes I use throughout this book are as follows:

(//) starts of overlap in talk are marked by a double oblique;

(.) pauses in talk where not timed but simply marked with one dot;

. . . denotes omitted material;

*italics* denotes words or particles said with emphasis;

[inaudible] has been used when the conversation could not be heard and transcribed;

comments in square brackets, such as [laughter], have been used to include non-verbal communication and events that help to give context to the conversation;

commas, full stops, question marks and exclamation marks have been added in a manner designed to improve the readability of the extracts while conveying their sense, as heard, as effectively as possible;

brief comments or simple acknowledgement tokens (for example, yes, mm) from me or others who are present have been placed in round brackets.

4 It is interesting to read Denise's remarks in relation to Young's (1990b: 192–193) work on breasts. In discussing breasts, Young draws on an Irigarayan metaphysics of fluids in order to problematise a Cartesian ontology of men's solidity and women's fluidity.

5 Young's claim has been derived from the work of Kristeva (1980: 237) who argues that 'the mother is simply the site of her proceedings'. For Kristeva (1980: 237) the maternal designates both a space and a series of functions and processes but it must not be confused with a subject, for maternity is a process without a subject – 'It happens, but I'm not there'. I do not subscribe to this notion that maternity or pregnancy is just something that 'happens' to a woman and that it does not involve any agency.

6 Kitzinger (1989: 86) writes:

> Looking after yourself in pregnancy, from the very first weeks, is probably more important for the welfare of your baby than anything else you can do. It ensures that you provide the best possible environment for the developing baby – and, equally important, it gives you the best chance of being healthy and full of vitality, ready for the birthday and the first stages of motherhood.

Even though Kitzinger assumes a positive, rather than an antagonistic, relationship between mother and fetus, she nevertheless prioritises the well-being of the fetus over the mother. The mother must look after herself for the sake of her fetus rather than for her own sake.

7 In the pamphlet, published by the New Zealand Department of Health, listeria is described as 'a common bacterium which is found in the soil, water, plants and in the droppings and faeces of animals and humans'. Sometimes listeria can cause a rare illness related to eating contaminated food. This infection is called listeriosis. Listeriosis is considered to be dangerous for pregnant women, as it can cause miscarriage and stillbirth.

## 4 MEN'S BODIES AND BATHROOMS

1 The label 'heterosexual' (like many other labels used to signify identity) is problematic. A binary division between heterosexual/homosexual needs to be problematised. While some men might identity as heterosexual (supporting heteronormativity by appearing 'straight', getting married and having children) they might also desire/engage in homosexual acts. Despite this instability of meaning, the term 'heterosexual' has currency in describing one facet of hegemony.

2 In New Zealand homes toilets are sometimes located in the bathroom and sometimes located in a separate room. Most homes only have one toilet/bathroom although new homes are often built with two (usually an en-suite to the 'master' bedroom, and a second toilet/bathroom for more general use). In my study eleven participants lived in a house containing one toilet and seven lived in a house containing two toilets. Fourteen participants lived in a house

containing one bathroom, three lived in a house containing two bathrooms, and one lived in a house containing three bathrooms. Half of the participants had a toilet in their bathroom while the other half had access to a separate toilet. Nearly all toilets in New Zealand are 'throne style' rather than 'squat style'. Sometimes toilets are referred to as 'lavatories' (although none of the participants used this term). Toilets are also referred to (more colloquially) as 'dunnies'.

3 In New Zealand the term 'flatting' refers to between approximately two and eight people (often aged 18–25) sharing a rented house, flat or apartment. Members of the household are referred to as 'flatmates'. 'Flatting', as a collective household arrangement, emerged in the early 1970s. At this time women usually 'flatted' with women and men 'flatted' with men. It did not take long, however, before men and women began to 'flat together' which was referred to as 'mixed flatting'. Sometimes friends move into a rental property to 'flat together' but often 'flatmates' are found through local advertising.

4 I am grateful to Robin Kearns for drawing my attention to this poem by W. H. Auden.

## 5 MANAGING MANAGERIAL BODIES

1 I am grateful to Melanie Hayes for drawing my attention to this photograph of the 'half-suit'.

2 It is also worth noting that while such dripping may result in a mark being produced on light coloured or lightweight shorts or trousers, the business suit made of heavy weight dark fabric with a jacket buttoned to cover the groin area, masks any such traces of insecure bodily boundaries.

3 Some participants claimed that winter conditions – freezing temperatures and darkness in the early morning and evening – make exercising, and finding the motivation to exercise, difficult. I suspect that several reasons underlie this claim. First, it feels unsafe for some people, especially women, to exercise outdoors in the dark. Second, it is often 'unpleasant' shifting from the warmth of 'inside spaces' to the coldness of 'outside spaces' for the purpose of exercise. It is also useful to note that in cooler climates people wear more clothes – the flesh is not exposed to the extent that it is in warmer climates.

## 6 SOME THOUGHTS ON CLOSE(T) SPACES

1 Despite this 'lack' in the academic geographical literature there are some accounts of domestic toilets/bathrooms in popular culture and fiction writing. For example, Philip Roth, in his memoir *Patrimony* (1991), tells the story of taking care of his dying father, Herman Roth, in the last year of his life. In the scene recounted below Herman loses control of his bowels, and his son Philip is compelled to clean up the mess.

> The shit was everywhere, smeared underfoot on the bathmat, running over the toilet bowl edge and, at the foot of the bowl, in a pile on the floor. It was splattered across the glass of the shower stall from which he'd just emerged, and the clothes discarded in the hallway were clotted with it . . . I saw that it was even on the tips of the bristles of my toothbrush hanging in the holder over the sink.
>
> (Roth 1991: 174)

Also, there are numerous magazines such as *New Zealand House and Garden*, and *New Zealand Bathroom Trends* that focus on the interiors and design of bathrooms. Magazines such as these tend to depict the spaces of bathrooms as completely devoid of people. The *New Zealand Bathroom Trends* magazine (volume 14, number 11) contains 121 colour pages – only five of these pages contain a person or people and these tend to be architects or designers rather than people using the bathroom. Barbara Hoffman (1996) in *Cosmopolitan* examines some of the problems of 'living together' in relation to sharing bathrooms. The newspaper *USA Today* published an article entitled '10 great places to . . . make a splash' (Clark 1999: 3D). Travel and Leisure Editor in Chief, Nancy Novogrod, checked out some of the most exclusive hotel bathrooms in Japan, Singapore, South Africa and the United States.

Toilets/bathrooms are also sometimes seen on television, especially in advertisements for cleaning products and sometimes in drama and comedy. For example, in an episode focusing on 'germaphobes', characters in the American television comedy series 'Seinfield' talk about disposable paper toilet seat covers and shower routines. Viewers see Jerry's bathroom, the showers at a YMCA (Young Men's Christian Association), Elaine's bathroom at work (that she shares with just one other colleague) and Kramer's shower. This episode culminates in Kramer installing a waterproof phone and a garbage disposal unit in his shower so that he can spend long periods in the shower carrying out everyday activities such as preparing meals.

2 It is worth noting that there is no clear mapping of the rational and irrational, mind and body, serious and fun binaries on to particular workplaces. Recently commentators (such as du Gay 1996, Leidner 1993, McDowell 1997) have commented on the blurring of the boundaries between work and leisure. For example, McDowell (1997: 139) explains that this blurring is found in the 'new built environment of the City [London] where, as the developers of the Broadgate Centre emphasised in their publicity material, the new buildings and the spaces around them are a "a total landscape of work and leisure"'. Shops and offices contain atria and lobbies that make them feel like large hotels. Workplaces and leisure places often blend under one roof. Downtown gyms, health clubs, swimming pools, tennis courts, casinos, hotels and gardens blend with shops and offices so that it is difficult to tell where one stops and the other starts.

3 The community of academic geographers in New Zealand is reasonably small. There are seven independent, public universities, six of which offer geography. In 1997 there were approximately 90 academic staff, 18 (20 percent) of whom are women. This is small compared with western countries such as the United Kingdom and the United States with far greater populations of academic geographers (Peace, Longhurst and Johnston 1997: 117).

# BIBLIOGRAPHY


Adler, S. and Brenner, J. (1992) 'Gender and space: lesbians and gay men in the city', *International Journal of Urban and Regional Research*, 16, 1: 24–34.

Ainley, R. (ed.) (1998) *New Frontiers of Space, Bodies and Gender*, London: Routledge.

Anderson, K. (1995) 'Culture and nature at the Adelaide Zoo: At the frontiers of "human" geography', *Transactions of the Institute of British Geographers*, 20: 275–294.

Angela, F. (1990) 'Confinement', in J. Rutherford (ed.) *Identity: Community, Culture and Difference*, London: Lawrence and Wishart: 72–87.

Auden, W. H. (1979) 'The geography of the house', in E. Mendelson (ed.) *W. H. Auden Selected Poems*, New York: Vintage Books.

*Baby on the Way*, (1994) June, Hastings, Infant Times.

Barnes, T. J. and Duncan, J. S. (eds) (1992) *Writing Worlds: Discourse, Text and Metaphor in the Representation of Landscape*, London: Routledge.

Barrett, M. (1980) *Women's Oppression Today: Problems in Marxist Feminist Analysis*, London: Verso.

Bell, D. (1991) 'Insignificant others: lesbian and gay geographies', *Area*, 23, 4: 323–329.

—— (1995) '[Screw]ing geography (censor's version)', editorial in *Environment and Planning D: Society and Space*, 13, 2: 127.

Bell, D., Binnie, J., Cream, J. and Valentine, G. (1994) 'All hyped up and no place to go', *Gender, Place and Culture: A Journal of Feminist Geography*, 1, 1: 31–47.

Bell, D. and Valentine, G. (1995) 'Introductions: orientations', in D. Bell and G. Valentine (eds) *Mapping Desire: Geographies of Sexualities*, London: Routledge: 1–27.

—— (eds) (1997) *Consuming Geographies: We Are What We Eat*, London: Routledge.

Berg, L. D. (1994) 'Masculinity, place, and a binary discourse of "theory" and "empirical investigation" in the human geography of Aotearoa/New Zealand', *Gender, Place and Culture: A Journal of Feminist Geography*, 1, 2: 245–260.

Berg, L. D. and Kearns, R. (1996) 'Naming as norming: "race", gender, and the identity politics of naming places in Aotearoa/New Zealand', *Environment and Planning D: Society and Space*, 46: 99–122.

—— (1998) 'America unlimited', *Environment and Planning D: Society and Space*, 16: 128–132.

Best, S. (1995) 'Sexualising space', in E. Grosz and E. Probyn (eds) *Sexy Bodies: The Strange Carnalities of Feminism*, New York: Routledge: 181–194.

Bhabha, H. K. (1990) 'The third space: interview with Homi Bhabha', in J. Rutherford (ed.) *Identity: Community, Culture, Difference,* London: Wishart and Lawrence: 207–221.

—— (1994) *The Location of Culture*, London: Routledge.

Binnie, J. (1997) 'Coming out of geography: towards a queer epistemology?', *Environment and Planning D: Society and Space*, 15: 223–237.

Birks, E. (1993) *Coping with 'Morning' Sickness*, Dunedin: University of Otago Press.

Blunt, A. and Rose, G. (eds) (1994) *Writing Women and Space: Colonial and Postcolonial Geographies*, New York: The Guilford Press.

Bondi, L. (1999) 'States on journeys: some remarks about human geography and psychotherapeutic practice', *Professional Geographer*, 51, 1: 11–24.

Bordo, S. (1986) 'The Cartesian masculinization of thought', *Signs*, 11: 239–256.

—— (1989) 'The body and the reproduction of femininity: a feminist appropriation of Foucault', in A. M. Jagger and S. Bordo (eds) *Gender/Body/Knowledge: Feminist Constructions of Being and Knowing*, New Brunswick: Rutgers University Press.

—— (1990) 'Reading the slender body', in M. Jacobus, E. Fox Keller and S. Shuttleworth (eds) *Body/Politics: Women and the Discourses of Science*, New York: Routledge.

—— (1993) *Unbearable Weight: Feminism, Western Culture, and the Body*, Berkeley: University of California Press.

Boston Women's Health Book Collective, The (1971) *Our Bodies Our Selves*, New York: Simon and Schuster.

Braidotti, R. (1989) 'The politics of ontological difference', in T. Brennan (ed.) *Between Feminism and Psychoanalysis,* London: Routledge.

—— (1991) *Patterns of Dissonance*, Cambridge: Polity Press.

Brophy, P. (1988) Story-boards in the film, *Salt, Saliva, Sperm and Sweat*, Melbourne.

Brown, M. (forthcoming) *Closet Spaces: Geographies of Metaphor from the Body to the Global*, London: Routledge.

Bullock, A., Stallybrass, O. and Trombley, S. (eds) (1988) *The Fontana Dictionary of Modern Thought*, London: Fontana Press.

Butler, J. (1990) *Gender Trouble: Feminism and the Subversion of Identity,* New York: Routledge.

—— (1993) *Bodies That Matter: On the Discursive Limits of Sex,* New York: Routledge.

—— (1997) *Excitable Speech: A Politics of the Performative*, New York: Routledge.

Butler, R. (1998) 'Rehabilitating the images of disabled youth', in T. Skelton and G. Valentine (eds) *Cool Places: Geographies of Youth Cultures*, London: Routledge: 83–100.

Butler, R. and Parr, H. (eds) (1999) *Mind and Body Spaces: Geographies of Illness, Impairment and Disability*, London: Routledge.

Buttimer, A. (1976) 'Grasping the dynamism of the lifeworld', *Annals of the Association of American Geographers*, 66: 277–292.

—— (1979) 'Reason, rationality and human creativity', *Geografiska Annaler*, 61B: 43–49.

Callard, F. J. (1998) 'The body in theory', *Environment and Planning D: Society and Space*, 16: 387–400.

Campbell, J. (1997) 'Creating vulnerability in natural hazards teaching', in E. Bliss (ed.) *Geo.Ed.97/Kaupapa Aro Whenua Geographical Education Conference*, University of Waikato, 6–9 July 1997: 134–137.

Census of Population and Dwellings (1991) *Waikato/Bay of Plenty Regional Report*, Wellington, Department of Statistics New Zealand.

Chavkin, W. (1992) 'Women and fetus: the social construction of conflict', in C. Feinman (ed.) *The Criminalization of a Woman's Body*, New York: Haworth Press: 193–202.

Chodorow, N. (1978) *The Reproduction of Mothering: Psychoanalysis and the Sociology of Gender*, Berkeley: University of California Press.

Chouinard, V. and Grant, A. (1996) 'On being not even anywhere near "the project"', in N. Duncan (ed.) *BodySpace*, London: Routledge: 170–193.

Christopherson, S. (1989) 'Flexibility in the US service economy and the emerging spatial division of labour', *Transactions of the Institute of British Geographers*, 14, 2: 131–143.

Clark, J. (1999) '10 great places to . . . make a splash', *USA Today*, March 26, 3D.

*Collins English Dictionary* (1979) Glasgow: William Collins.

Collins, P. Hill (1990) *Black Feminist Thought: Knowledge, Consciousness and the Politics of Empowerment*, London: Harper Collins.

Colomina, B. (1992) *Sexuality and Space*, New York: Princeton Architectural Press.

*Compact Oxford English Dictionary* (second edition) (1991) Oxford: Clarendon Press.

Connell, R. W. (1995) *Masculinities*, Berkeley: University of California Press.

Cook, G. (1990) 'Transcribing infinity: problems of context presentation', *Journal of Pragmatics*, 14: 1–24.

Crawford, R. (1984) 'A cultural account of "health" control, release and the social body', in J. McKinlay (ed.) *Issues in the Political Economy of Health Care*, New York: Tavistock: 60–103.

Crawford, S. (1987) '"One's nerves and courage are in very different order out in New Zealand": recreational and sporting opportunities for women in a remote colonial setting', in J. A. Mangan and R. J. Park (eds) *From 'Fair Sex' to Feminism*, London and Totowa: Frank Cass.

Cresswell, T. (1997) 'Weeds, plaques, and bodily secretions: a geography of interpretation of metaphors of displacement', *Annals of the Association of American Geographers*, 87, 2: 330–345.

—— (1999) 'Embodiment, power and the politics of mobility: the case of female tramps and hobos', *Transactions of the Institute of British Geographers*, 24: 175–192.

Davidson, J. and Smith, M. (1999) 'Wittgenstein and Irigaray: gender and philosophy in a language (game) of difference', *Hypatia*, 14, 2: 72–96.

Davis, K. (ed.) (1997) *Embodied Practices: Feminist Perspectives on the Body*, London: Sage.

Dawson, J. B. (1953) *The Expectant Mother*, Wellington: Whitcombe and Tombs.

Dawson, R. (1983) *Customs of Childbirth: Migrant Women From Twelve Different Cultures Speak of their Own Experiences and Customs*, Wellington: Wellington Multicultural Educational Resource Centre.

De Beauvoir, S. (1953) *The Second Sex*, trans. H. M. Parshley, London: Cape.

De Lauretis, T. (1986) 'Feminist studies/critical studies: issues, terms and contexts', in De Lauretis (ed.) *Feminist Studies/Critical Studies*, London: Macmillan, 1–19.

Dear, M. and Scott, A. J. (eds) (1981) *Urbanization and Urban Planning in Capitalist Society*, London and New York: Methuen.

Deem, H. and Fitzgibbon, N. P. (1953) *Modern Mothercraft: A Guide to Parents*, Official Handbook Royal New Zealand Society for the Health of Women and Children (Incorporated) Plunket Society: Dunedin.

Deleuze, G. and Guattari, F. (1983) *Anti-Oedipus: Capitalism and Schizophrenia*, trans. R. Hurley, M. Seem and H. Lane, Minneapolis: University of Minnesota Press.

—— (1986) *Nomadology: the War Machine*, trans. B. Massumi, New York: Semiotext(e).

Dinnerstein, D. (1976) *The Mermaid and the Minotaur: Sexual Arrangements and Human Malaise*, New York: Harper and Row.

Dixon, G. (1999) 'Sex and filth', *New Zealand Herald*, J5.

Domosh, M. (1997) 'Geography and gender: the personal and the political', *Progress in Human Geography*, 21, 1: 81–87.

Dorn, M. (1998) 'Beyond nomadism: the travel narratives of a "cripple"', in S. Pile and H. Nast (eds) *Places Through the Body*, London, Routledge: 183–206

Dorn, M. and Laws, G. (1994) 'Social theory, body politics and medical geography: extending Kearn's invitation', *Professional Geographer*, 46, 1: 106–110.

Douglas, M. (1966) *Purity and Danger: An Analysis of Concepts of Pollution and Taboo*, London: Routledge.

—— (1975) *Implicit Meanings: Essays in Anthropology*, London: Routledge.

Downs, R. M. and Stea, D. (1977) *Maps in Minds: Reflections on Cognitive Mapping*, New York: Harper and Row.

Du Gay, P. (1996) *Consumption and Identity at Work*, London: Sage.

Duncan, N. (ed.) (1996) *BodySpace: Destabilizing Geographies of Gender and Sexuality*, London: Routledge.

Dyck, I. (1995) 'Hidden geographies: the changing lifeworlds of women with multiple sclerosis', *Soc. Sci. Med.*, 40, 3: 307–320.

—— (1999): 'Body troubles: women, the workplace and negotiations of a disabled identity', in R. Butler, and H. Parr (eds) *Mind and Body Spaces, New Geographies of Illness, Impairment and Disability*, London: Routledge.

Edinburgh Facts and Figures (1995) Available HTTP: <http://www.efr.hw.ac.uk/EDC/facts-and-figures95/contents.html> (accessed 26 November 1999).

Edwards, J. and McKie, L. (1997) 'Women's public toilets: a serious issue for the body politic', in K. Davis (ed.) *Embodied Practices: Feminist Perspectives on the Body*, London: Sage, 135–149.

Elder, G. (1995) 'Of moffies, kaffirs and perverts: male homosexuality and the discourse of moral order in the apartheid state', in D. Bell and G. Valentine (eds) *Mapping Desire: Geographies of Sexualities*, London: Routledge: 56–65.

Entrikin, J. N. (1976) 'Contemporary humanism in geography', *Annals of the Association of American Geographers*, 66: 615–632.

Fairclough, N. (1992) *Discourse and Social Change*, Cambridge: Polity Press.

Featherstone, M. (1983) 'The body in consumer culture', *Theory, Culture & Society*, 1, 2: 18–33.

Foord, J. and Gregson, N. (1986) 'Patriarchy: towards a reconceptualisation', *Antipode*, 18, 2: 186–211.

Foster, S. L. (ed.) (1996) *Corporealities – Dancing Knowledge, Culture and Power*, London: Routledge.

Foucault, M. (1970) *The Order of Things: An Archaeology of the Human Sciences*, London: Tavistock.

—— (1979) *Discipline and Punish: The Birth of the Prison*, trans. A. Sheridan, London, Penguin.

—— (1980) *The History of Sexuality, Volume 1: An Introduction*, trans. R. Hurley, New York, Vintage/Random House.

—— (1985) *The Use of Pleasure, Volume 2 of the History of Sexuality*, trans. R. Hurley, New York: Pantheon.

—— (1986) *The Care of the Self, Volume 3 of the History of Sexuality*, trans. R. Hurley, New York: Pantheon.

Frank, A. W. (1990) 'Bringing bodies back in: a decade review', *Theory, Culture & Society*, 7: 131–162.

—— (1991) 'For a sociology of the body: an analytical review', in M. Featherstone, M. Hepworth and B. S. Turner (eds) *The Body: Social Process and Cultural Theory*, London: Sage, 36–102.

Frye, M. (1983) *The Politics of Reality: Essays in Feminist Theory*, Trumansburg: Crossing Press.

Fujita, K. (1991) 'Women workers and flexible specialization: the case of Tokyo', *Economy and Society*, 20, 3: 260–282.

Gallop, J. (1988) *Thinking Through the Body*, New York: Columbia University Press.

Garner, P. (1983) *Better Living Catalog*, London: Sidgwick and Jackson: 30.

Gatens, M. (1988) 'Towards a feminist philosophy of the body', in B. Caine, E. Grosz and M. de Lepervanche (eds) *Crossing Boundaries: Feminisms and Critiques of Knowledges*, Sydney: Allen and Unwin.

—— (1991a) 'Corporeal representation in/and the body politic', in R. Diprose and R. Ferrell (eds) *Cartographies: Poststructuralism and the Mapping of Bodies and Spaces*, Sydney: Allen and Unwin: 79–87.

—— (1991b) 'A critique of the sex/gender distinction', in S. Gunew *A Reader in Feminist Knowledges*, New York: Routledge.

—— (1996) *Imaginary Bodies: Ethics, Power and Corporeality*, London: Routledge.

Gibson-Graham, J. K. (1996) *The End of Capitalism (As We Knew It): A Feminist Critique of Political Economy*, Cambridge: Blackwell.

—— (1997) 'Postmodern becomings: from rape space to pregnant space', in G. B. Benko and U. Strohmayer (eds) *Space and Social Theory: Geographical Interpretations of Postmodernity*, Oxford: Blackwell: 306–323.

Gould, J. (1958) *Will My Baby Be Born Normal?* Public Affairs Committee, United States.

Gould, P. and R. White (1974) *Mental Maps*, London: Penguin.

Greer, G. (1970) *The Female Eunuch*, London: MacGibbon and Kee.

Grosz, E. (1988) 'Desire, the body and recent French feminisms', *Intervention*, 21/22: 28–33.

—— (1989) *Sexual Subversions: Three French Feminists*, Sydney: Allen and Unwin.

—— (1990) 'The body of signification', in J. Fletcher and A. Benjamin (eds) *Abjection, Melancholia and Love: The Work of Julia Kristeva*, London: Routledge: 80–103.

—— (1992) 'Bodies-cities', in B. Colomina (ed.) *Sexuality and Space*, New York: Princeton Architectural Press.

—— (1993) 'Bodies and knowledges: feminism and the crisis of reason', in L. Alcoff and E. Potter (eds) *Feminist Epistemologies,* New York: Routledge: 187–215.

—— (1994a) *Volatile Bodies: Toward a Corporeal Feminism*, St Leonards: Allen and Unwin.

—— (1994b) 'Women, *chora*, dwelling', in S. Watson and K. Gibson (eds) *Postmodern Cities and Spaces*, Oxford: Blackwell: 47–58.

—— (1995) *Space, Time and Perversion*, London: Routledge.

Grosz, E. and Probyn, E. (eds) (1995) *Sexy Bodies: The Strange Carnalities of Feminism*, New York: Routledge.

Hanna, J. L. (1931) *Modern New Zealand Homes*, Auckland: Home and Garden Services.

Haraway, D. (1990) 'A manifesto for cyborgs', in L. Nicholson (ed.) *Feminism/Postmodernism,* New York: Routledge: 190–233.

—— (1991) *Simians, Cyborgs, and Women: The Reinvention of Nature*, London: Free Association Books.

Harré, R. (1991) *Physical Being*, Oxford: Blackwell.

Harvey, D. (1973) *Social Justice and the City*, London: Edward Arnold.

—— (1989) *The Condition of Postmodernity*, London: Blackwell.

Hayden, D. (1976) *Seven American Utopias: The Architecture of Communitarian Socialism, 1790–1975*, Cambridge, MA: MIT Press.

—— (1978) 'Two utopian feminists and their campaign for kitchenless houses', *Signs*, 4, 2: 274–290.

Hodder, B. W. and Lee, R. (1974) *Economic Geography*, London, Methuen.

Hodge, S. (1995) '"No fags out there": gay men, identity and suburbia', *Journal of Interdisciplinary Gender Studies*, 1, 1: 41–48.

Hoffman, B. (1996) 'Living together . . .', *Cosmopolitan*, 1 December, 221 (6): 116.

hooks, b. (1990) *Yearning: Race, Gender and Cultural Politics,* Boston: South End Press.

Hughes, E. C. (1964) 'Can a mother's illness harm her unborn baby?', *Redbook*, July: 22.

Irigaray, L. (1985) *This Sex Which is Not One*, trans. C. Porter with C. Burke, New York: Cornell University Press.

—— (1993) *je, tu, nous: Toward a Culture of Difference*, trans. A. Martin, London: Routledge.

Jackson, P. (1991) 'The cultural politics of masculinity: towards a social geography', *Transactions of the Institute of British Geographers*, 16: 199–213.

—— (1993) 'Towards a cultural politics of consumption', in J. Bird, B. Curtis, T. Putnam, G. Robertson and L. Tickner (eds) *Mapping the Futures: Local Cultures, Global Change*, London: Routledge.

—— (1994) 'Black male: Advertising and the cultural politics of masculinity', *Gender, Place and Culture: A Journal of Feminist Geography*, 1, 1: 49–59.

Jackson, P., Stevenson, N. and Brooks, K. (1999) 'Making sense of men's lifestyle magazines', *Environment and Planning D: Society and Space*, 17: 353–368

Jaggar, A. M. and Bordo, S. R. (1989) *Gender/Body/Knowledge: Feminist Reconstructions of Being and Knowing*, New Brunswick: Rutgers University Press.

Jarosz, L. (1992) 'Constructing the dark continent metaphors as geographic representation of Africa', *Geografiska Annaler*, 74B: 105–115.

Jay, N. (1981) 'Gender and dichotomy', *Feminist Studies*, 7: 38–56.

Jefferson, G. (1985) 'An exercise in the transcription and analysis of laughter', in T. van Dijk (ed.) *Handbook of Discourse Analysis*, 3, London: Academic Press.

Johnson, L. C. (1989) 'Embodying geography – some implications of considering the sexed body in space', *New Zealand Geographical Society Proceedings of the 15th New Zealand Geography Conference,* Dunedin, August: 134–138.

—— (1990) 'New courses for a gendered geography: Teaching feminist geography at the University of Waikato', *Australian Geographical Studies*, 28, (1): 16-27.

—— (1992) 'Housing desire: a feminist geography of suburban housing', *Refractory Girl*, 42: 40–47.

—— (1994a) 'What future for feminist geography?', *Gender, Place and Culture: A Journal of Feminist Geography*, 1, 1: 103–114.

—— (1994b) 'Colonising the suburban frontier: place-making on Melbourne's urban fringe', in K. Gibson and S. Watson (eds) *Metropolis Now*, Leichhardt: Pluto.

Johnston, L. (1997) 'Queen(s') Street or Ponsonby Poofters? Embodied HERO parade sites', *New Zealand Geographer*, 53, 2: 29–33.

Johnston, L. and G. Valentine, (1995) 'Wherever I lay my girlfriend, that's my home', in G. Bell and D. Valentine (eds) *Mapping Desire: Geographies of Sexualities*, London: Routledge: 99–113.

Johnston, R. (1984) 'Marxist political economy, the state and political geography', *Progress in Human Geography*, 8: 473–492.

—— (1986) *Philosophy and Human Geography: An Introduction to Human Geography*, London: Edward Arnold.

Johnston, R, Gregory, D. and Smith D. M. (eds) (1986) *The Dictionary of Human Geography*, Oxford: Basil Blackwell.

Jones, D. (1992) 'Gender at work: managerial women's clothing as performance and representation', paper presented at the Canterbury Critical Theory Group Conference on Forms, Formations and Malformations, Christchurch: 13–14 November.

Jones, J. P., Nast, H. J. and Roberts, S. M. (eds) (1997) *Thresholds in Feminist Geography: Difference, Methodology, Representation*, Oxford: Rowman and Littlefield.

*Journal of Geography in Higher Education* (1999) 'JGHE Symposium: teaching sexualities in geography', 23, 1: 77–123.

Kantor, R. M. (1989) *When Giants Learn to Dance*, Simon and Schuster: New York.

Katz, C. (1992) 'All the world is staged: intellectuals and the projects of ethnography', *Environment and Planning D: Society and Space*, 10: 495–510.

—— (1994) 'Playing the field: questions of fieldwork in geography', *The Professional Geographer,* 46: 67–72.

Keogh, P. (1992) 'Public sex: spaces, acts, identities', Proceedings of the Sexuality and Space Network, One Day Seminar: 'Lesbian and Gay Geographies?'

Kerr, A. (1967) 'Protecting the unborn baby', *McCall's*, July: 48.

Kira, A. (1966) *The Bathroom: Criteria for Design*, New York: Bantam Books.

Kirby, V. (1989) '"Corporeographies", in *Traveling Theories Traveling Theorists*' in J. Clifford and V. Dhareshwar (eds) (Group for the Critical Study of Colonial Discourse and the Center for Cultural Studies, University of California at Santa Cruz) *Inscriptions*, 5: 103–119.

—— (1991) '*Corpus delicti:* the body at the scene of writing', in R. Diprose and R. Ferrell (eds) *Cartographies: Poststructuralism and the Mapping of Bodies and Spaces*, Sydney: Allen and Unwin: 88–100.

—— (1992) *Addressing Essentialism Differently . . . Some Thoughts on the Corpo-real*, Occasional Paper Series, No. 4, University of Waikato, Department of Women's Studies.

—— (1997) *Telling Flesh: The Substance of the Corporeal*, New York: Routledge.

Kitzinger, S. (1989) *Pregnancy and Childbirth*, London: Doubleday.

Knopp, L. (1990a) 'Social consequences of homosexuality', *Geographical Magazine*, 62, 5: 20–25.

—— (1990b) 'Some theoretical implications of gay involvement in the urban land market', *Political Geography Quarterly*, 9: 337–352.

—— (1992) Sexuality and the spatial dynamics of capitalism, *Environment and Planning D: Society and Space*, 10, 6: 651–670.

Kristeva, J. (1980) *Desire in Language: A Semiotic Approach to Literature and Art*, ed. Leon S. Roudiez, New York: Columbia University Press.

—— (1982) *Powers of Horror: An Essay on Abjection*, trans. Leon S. Roudiez, New York: Columbia University Press.

Lacan J. (1981) *Le Séminaire*, Book 20, *Encore*, Paris: Seuil.

Law, R., Campbell, H. and Dolan, J. (eds) (1999) *Masculinities in Aotearoa/New Zealand*, Palmerston North: Dunmore Press.

Law, R., Cooper, A. Malthus, J. and Wood, P. (1999) 'Bodies, sites, and citizens: the politics of public toilets', *Proceedings of the Southern Regional Conference of the International Geographical Union Commission on Gender*, University of Otago, 8–11 February, compiled by R. Longhurst and R. Peace.

Laws, G. (1997) 'Women's life courses, spatial mobility, and state policies', in J. P. Jones III, H. J. Nast and S. M. Roberts (eds) *Thresholds in Feminist Geography: Difference, Methodology, Representation*, Oxford: Rowman and Littlefield: 47–64.

Le Doeuff, M. (1987) 'Women in philosophy', in T. Moi (ed.) *French Feminist Thought: A Reader*, Oxford: Blackwell.

—— (1991) *Hipparchia's Choice: An Essay Concerning Women, Philosophy, etc.*, Oxford: Blackwell.

Lechte, J. (1993) '(Not) belonging in postmodern space', in S. Watson and K. Gibson (eds) *Postmodern Cities Conference Proceedings*, 14–16 April, Department of Urban and Regional Planning, University of Sydney.

Leidner, R. (1993) *Fast Food, Fast Talk; Service Work and the Routinization of Everyday Life*, University of California Press: Berkley.

Ley, D. and Samuels, M. S. (1978) 'Introduction: contexts of modern humanism in geography', in D. Ley and M. S. Samuels (eds) *Humanistic Geography: Prospects and Problems*, London, Croom Helm: 1–17.

Lloyd, G. (1993) *The Man of Reason: 'Male' and 'Female' in Western Philosophy*, London: Routledge.

Longhurst, R. (1994a) Reflections on and a vision for feminist geography, *New Zealand Geographer*, 50, 1: 14–19.

—— (1994b) 'Review of P. Rodaway 1994 Sensuous Geographies: Body, Sense and Place

(Routledge)', in *Tijdschrift Voor Economische en Sociale Geographie*, 87, 5: 465–466.
—— (1995a) 'Discursive constraints on pregnant women's participation in sport', *New Zealand Geographer*, 51, 1: 13–15.
—— (1995b) 'Geography and the body', *Gender, Place and Culture: A Journal of Feminist Geography*, 2, 1: 97–105.
—— (1996a) 'Refocusing groups: pregnant women's geographical experiences of Hamilton, New Zealand/Aotearoa', *Area*, 28, 2: 143–149.
—— (1996b) 'Geographies that matter: pregnant bodies in public places', unpublished D. Phil. thesis, University of Waikato.
—— (1997) '(Dis)embodied geographies', *Progress in Human Geography*, 21, 4: 486–501.
—— (1998) '(Re)presenting shopping centres and bodies: questions of pregnancy', in R. Ainley (ed.) *New Frontiers of Space, Bodies and Gender*, London: Routledge: 20–34.
—— (forthcoming 2000) '"Corporeographies" of pregnancy: "bikini babes"', *Environment and Planning D: Society and Space*, 18.
Lynch, K. (1960) *The Image of the City*, Cambridge: MIT Press.
McClintock, A. (1995) *Imperial Leather: Race, Gender and Sexuality in the Colonial Conquest*, New York: Routledge.
Macdonald, J. (1992) 'The hidden bits', unpublished paper, Department of Anthropology, University of Waikato.
McDowell, L. (1983) 'Towards an understanding of the gender division of urban space', *Environment and Planning D: Society and Space*, 1: 59–72.
—— (1993) 'Space, place and gender relations: part II identity, difference, feminist geometries and geographies', *Progress in Human Geography*, 17, 3: 305–318.
—— (1995) 'Body work: heterosexual gender performances in city workplaces', in D. Bell and G. Valentine (eds) *Mapping Desire: Geographies of Sexualities,* London: Routledge: 75–95
—— (1997) *Capital Culture: Gender at Work*, Oxford: Blackwell.
—— (1999) *Gender, Identity and Place: Understanding Feminist Geographies*, Polity Press: Cambridge.
McDowell, L. and Court, G. (1994) 'Performing work: bodily representations in merchant banks', *Environment and Planning D: Society and Space*, 12: 727–750.
MacKenzie, S. (1984) 'Editorial introduction', *Antipode*, 16, 3: 3–10.
—— (1987) 'Neglected spaces in peripheral places: homeworkers and the creation of a new economic centre', *Cahiers di Geographic du Quebec*, 31, 83: 247–260.
McKinnon, M. (ed.) with Bradley, B. and Kirkpatrick, R. (1997) *New Zealand Historical Atlas Ko Papatuanuku e Takoto Nei*, David Bateman in association with Historical Branch, Department of Internal Affairs: Auckland.
McNee, B. (1984) 'If you are squeamish', *East Lakes Geographer*, 19: 16–27.
Malbon, B. (1998) 'Clubbing: consumption, identity and the spatial practices of every-night life', in T. Skelton and G. Valentine (eds) *Cool Places: Geographies of Youth Cultures*, London: Routledge: 266–286.
Marcus, S. (1992) 'Fighting bodies, fighting words: a theory and politics of rape prevention', in J. Butler and J. Scott (eds) *Feminists Theorize the Political*, London: Routledge: 385–403.
—— (1993) 'Placing Rosemary's baby', *Differences: A Journal of Feminist Cultural Studies*, 5, 3: 121–153.
Margo, J. (1996) *Man Maintenance*, Ringwood, Australia: Penguin.
Martin, E. (1994) *Flexible Bodies: The Role of Immunity in American Culture From the Days of Polio to the Age of AIDS*, Boston: Beacon Press.
Massey, D. (1998) 'Blurring the binaries? High tech in Cambridge', in R. Ainley (ed.) *New Frontiers of Space, Bodies and Gender*, London: Routledge: 157–175.

Matrix (1984) *Making Space: Women and the Man Made Environment*, London: Pluto Press.

*Metro* (1993) 'Last word', Auckland: 138.

Miles, M. B. and Huberman, A. M. (1994) *Qualitative Data Analysis: An Expanded Sourcebook*, Thousand Oaks: Sage.

Millet, K. (1970) *Sexual Politics*, New York: Doubleday.

Mohanram, R. (1999) *Black Body: Women, Colonialism and Space*, St Leonards NSW: Allen and Unwin.

Molloy, J. (1977) *The Woman's Dress for Success Book*, New York: Warner Books.

Morales, R. (1983) 'The other heritage', in C. Moraga and G. Anzaldúa (eds) *This Bridge Called My Back: Writing by Radical Women of Color*, New York: Kitchen Press, 107–108.

Morgan, D. (1993) 'You too can have a body like mine: reflections on the male body and masculinities', in S. Scott and D. H. J. Morgan, (eds) *Body Matters*, London: The Falmer Press: 69–88.

Moss, P. and Dyck, I. (1999) 'Body, corporeal space, and legitimating chronic illness: women diagnosed with M.E.', *Antipode*, 31, 4: 372–397.

Munt, S. R. (1998) *Heroic Desire: Lesbian Identity and Cultural Space*, London: Cassell: 76–80.

Nagar, R. (1997) 'Exploring methodological borderlands through oral narratives', in J. P. Jones, H. J. Nast and S. M. Roberts (eds) *Thresholds in Feminist Geography: Difference, Methodology, Representation*, Oxford: Rowman and Littlefield: 203–224.

Namaste, K. (1996) 'Genderbasing: sexuality, gender, and the regulation of public space', *Environment and Planning D: Society and Space*, 14: 221–240.

Nash, C. (1994) 'Remapping the body/land: new cartographies of identity, gender, and landscape in Ireland', in A. Blunt and G. Rose (eds) *Writing Women and Space: Colonial and Postcolonial Geographies*, New York: The Guilford Press: 227–250.

Nast, H. J. and Pile, S. (eds) (1998) *Places Through the Body*, London: Routledge.

*New Zealand Bathroom, Trends* (1999) 14: 11, 121.

Oakley, A. (1972) *Sex, Gender and Society,* London: Maurice Temple Smith.

Ochs, E. (1979) 'Transcription as theory', in E. Ochs and B. Schieffelin (eds) *Developmental Pragmatics*, New York: Academic Press.

Oliver, K. (1993) *Reading Kristeva: Unravelling the Double Bind*, Indiana University Press: Bloomington.

*Oxford English Dictionary (The Compact)* (1991) Oxford: Clarendon Press.

Peace, R. (1997) 'CAQDAS/NUD.IST: computer assisted qualitative data analysis software/non-numerical, unstructured data. Indexing, searching and theorising – a geographical perspective', in E. Bliss (ed.) *Conference Proceedings, Second Joint Conference, Institute of Australian Geographers and New Zealand Geographical Society*, University of Tasmania, 28–31 January: 382–385.

—— (1999) Surface tension: place/poverty/policy. 'From "poverty" to "social exclusion": implications of discursive shifts in European Policy 1975–1999', unpublished PhD. thesis, University of Waikato.

Peace, R., Longhurst, R. and Johnston, L. (1997) 'Producing feminist geography "down under"', *Gender, Place and Culture: A Journal of Feminist Geography*, 4, 1: 115–119.

Phillips, J. (1987) *A Man's Country? The Image of the Pakeha Male – A History*, Auckland: Penguin Books.

Pile, S. (1996) *The Body and the City: Psychoanalysis, Space and Subjectivity*, London: Routledge.

—— (1998) 'Freud, dreams and imaginative geographies', in M. Elliott (ed.) *Freud 2000*, Carlton, Victoria: Melbourne University Press.

Pile, S. and Thrift, N. (eds) 1995: *Mapping the Subject: Geographies of Cultural Transformation*, London: Routledge.

Pilgrim, J. (1993) 'Language and lack: Kiki Smith and representation of the naked pregnant body', unpublished paper, University of Western Sydney, Nepean.

Porteous, D. (1991) 'Hamilton East: interpretations of house styles and infilling the one acre section', unpublished M.SocSc. thesis, University of Waikato.

Pratt, M. B. (1984) 'Identity: skin, blood, heart', in E. Burkin, M. B Pratt and B. Smith (eds) *Yours in Struggle: Three Feminist Perspectives on Anti Semitism and Racism*, Bloomington: Long Haul Press.

Probyn, E. (1993) *Sexing the Self: Gendered Positions in Cultural Studies*, London: Routledge.

Pulido, L. (1997) 'Community, place, and identity', in J. P. Jones, H. J. Nast and S. M. Roberts (eds) *Thresholds in Feminist Geography: Difference, Methodology, Representation*, Oxford: Rowman and Littlefield: 11–28.

Pulsipher, L. M. (1997) 'For whom shall we write? What voice shall we use? Which story shall we tell?', in J. P. Jones, H. J. Nast and S. M. Roberts (eds) *Thresholds in Feminist Geography: Difference, Methodology, Representation*, Oxford: Rowman and Littlefield: 319–337.

Radcliffe, S. A. (1999) 'Embodying national identities: *mestizo* men and white women in Ecuadorian racial–national imaginaries', *Transactions of the Institute of British Geographers*, 24: 213–225.

Randell, M. (1945) *Training for Childbirth: From the Mother's Point of View*, London: J. and A. Churchill.

Reimer, S. (1994) 'Flexibility and the gender division of labour: The restructuring of public sector employment in British Columbia', *Area*, 26, 4: 351–358.

Relph, E. (1976) *Place and Placenesslesss*, London: Pion.

—— (1981) *Rational Landscapes and Humanistic Geography*, London: Croom Helm.

Rich, A. (1986) 'Notes towards a politics of location', in A. Rich, *Blood, Bread and Poetry: Selected Prose 1979–1985*, New York: W. W. Norton and Company.

Richard, B. and Kruger, H. H. (1998) 'Ravers' paradise?: German youth cultures in the 1990s', in T. Skelton and G. Valentine (eds) (1998) *Cool Places: Geographies of Youth Cultures*, London: Routledge: 161–174.

Rodaway, P. (1994) *Sensuous Geographies: Body, Sense and Place*, London: Routledge.

Rose, G. (1991) 'On being ambivalent: women and feminisms in geography', in C. Philo (ed.) *New Words, New Worlds: Reconceptualising Social and Cultural Geography,* Conference Proceedings, Department of Geography, University of Edinburgh, 10–12 September.

—— (1993a) *Feminism and Geography: The Limits of Geographical Knowledge*, Cambridge: Polity Press.

—— (1993b) 'Speculations on what the future holds in store', *Environment and Planning A: Anniversary Issue*, 26–29.

—— (1995) 'Geography and gender, cartographies and corporealities', *Progress in Human Geography*, 19, 4: 544–48.

—— (1997) 'Situating knowledges: positionality, reflexivities and other tactics', *Progress in Human Geography*, 21, 3: 305–320.

Roth, P. (1991) *Patrimony: A True Story*, New York: Simon and Schulster.

Rothman, B. K. (1979) 'Women, health and medicine', in J. Freeman (ed.) *Women: A Feminist Perspective*, California: Mayfield Publishing: 27–40.

Rowles, G. D. (1978a) 'Reflections on experiential field work', in D. Ley and M. Samuels (eds) *Humanistic Geography: Prospects and Problems,* Chicago: Maaroufa Press.

—— (1978b) *The Prisoners of Space? Exploring the Geographical Experience of Older People*, Boulder: Westview Press.

Saarinen, T. F. (1976) 'The bathroom: an example of industrial design', in *Environmental Planning: Perception and Behaviour*, Boston: Houghton Mifflin: 21-24.

Salmond, J. (1986) *Old New Zealand Houses 1800–1940*, Auckland: Reed Methuen.

Sartre, J. P. (1956) *Being and Nothingness*, trans. H. E. Barnes, New York: Philosophical Library.

Scott, S. and Morgan, D. (eds) (1993) *Body Matters: Essays on the Sociology of the Body*, London: Falmer Press.

Sedgwick, E. K. (1990) *Epistemology of the Closet*, Berkeley: University of California Press.

Sharpe, S. (1999) 'Bodily speaking: spaces and experiences of childbirth', in E. Teather (ed.) *Embodied Geographies: Spaces, Bodies and Rites of Passage*, London: Routledge.

Shildrick, M. (1997) *Leaky Bodies and Boundaries: Feminism, Postmodernism and (Bio)ethics*, London: Routledge.

Shilling, C. (1993) *The Body and Social Theory*, London: Sage.

Short, J. (1985) 'Human geography and Marxism', in Z. Baranski and J. Short (eds) *Developing Contemporary Marxism*, London: Macmillan: 165–195.

Sibley, D. (1995) *Geographies of Exclusion: Society and Difference in the West*, London: Routledge.

Silverman Van Buren, J. (1989) *The Modernist Madonna: Semiotics of the Maternal Metaphor*, Bloomington and Indianapolis: Indiana University Press.

Skelton, T. and Valentine, G. (eds) (1998) *Cool Places: Geographies of Youth Cultures*, London: Routledge.

Smith, N. (1984) *Uneven Development: Nature, Capital and the Production of Space*, Oxford: Basil Blackwell.

Smith, G. and Winchester, H., (1998) 'Negotiating space: alternative masculinities at the work/home boundary', *Australian Geographer*, 29, 3: 327–339.

Soja, E. W. (1996) *Thirdspace: Journey to Los Angeles and Other Real-and-Imagined Places*, Cambridge, MA: Blackwell.

Soja, E. and Hooper, B. (1993) 'The spaces that difference makes: some notes on the geographical margins of the new cultural politics', in S. Pile and M. Keith (eds) *Place and the Politics of Identity*, London: Routledge: 183–205

Spain, D. (1992) *Gendered Space*, Chapel Hill: University of North Carolina Press.

Spoonley, P. (1993) *Racism and Ethnicity*, Auckland: Oxford University Press.

Stallybrass, P. and White, A. (1986) *The Politics and Poetics of Transgression*, London: Methuen.

Statistics New Zealand, Te Tari Tatau (1996) *New Zealand Official 1996 Year Book*, 99th Edition: Auckland.

Stoller, R. (1968) *Sex and Gender,* London: Hogarth.

Teather, E. (ed.) (1999) *Embodied Geographies: Spaces, Bodies and Rites of Passage*, London: Routledge.

Theweleit, K. (1987a) *Male Fantasies I: Women, Floods, Bodies, History*, Cambridge: Polity Press.

—— (1987b) *Male Fantasies II: Male Bodies: Psychoanalyzing the White Terror*, Cambridge: Polity Press.

Tuan, Yi-Fu. (1974) *Topophilia: A Study of Environmental Perception, Attitudes and Values*, Englewood Cliffs: Prentice-Hall.

—— (1975) 'Images and mental maps', *Annals of the Association of American Geographers*, 65, 2: 205–213.

—— (1976) 'Humanistic geography', *Annals of the Association of American Geographers*, 66: 266–276.

—— (1979) *Landscapes of Fear*, Oxford: Basil Blackwell.

Turner, B. (1984) *The Body and Society: Explorations in Social Theory*, Oxford: Basil Blackwell.

Valentine, G. (1993) '(Hetero)sexing space: lesbian perception and experiences of everyday spaces', *Environment and Planning D: Society and Space*, 11: 395–413.

—— (1997) 'Ode to a geography teacher: sexuality and the classroom', *Journal of Geography in Higher Education*, 21, 3: 417–422.

—— (1998) 'Food and the production of the civilised street', in N. R. Fyfe (ed.) *Images of the Street: Planning, Identity and Control in Public Space*, London: Routledge: 192–204.

*Waikato Business News* (1996) advertisement for OCOM Office Communication Limited, 1 July.

*Waikato Weekender*, (1991) 'Emotional changes are a normal part of pregnancy', 9 March: 14.

Weisman, L. K.(1992) *Discrimination by Design: A Feminist Critique of the Man-Made Environment*, Urbana: University of Illinois Press.

Weiss, G. (1999) *Body Images: Embodiment as Intercorporeality*, New York: Routledge.

Whelan, E. M. (1982) *Eating Right: Before, During and After Pregnancy*, Wauwatosa: American Baby Books.

Wolff, J. (1990) *Feminine Sentences: Essays on Women and Culture*, Cambridge: Polity Press.

Women and Geography Study Group of the IBG (1984) *Geography and Gender: An Introduction to Feminist Geography*, London: Century Hutchinson.

Woods, G. (1995) 'Fantasy islands: popular topographies of marooned masculinity', in D. Bell and G. Valentine (eds) *Mapping Desire: Geographies of Sexualities*, London: Routledge: 126–148.

Wright, E (ed.) (1992) *Feminism and Psychoanalysis: A Critical Dictionary*, Oxford: Blackwell Reference.

Young, I. (1990a) 'The scaling of bodies and the politics of identity', in *Justice and the Politics of Difference*, Princeton: Princeton University Press: 122–155.

—— (1990b) *Throwing Like a Girl and Other Essays in Feminist Philosophy and Social Thought*, Indianapolis: Indiana University Press: 160–174.

*Your Pregnancy To Haputanga*, (1991) Department of Health New Zealand.

# INDEX